CATALOGUE OF EUROPEAN AMBERS

Catalogue of

EUROPEAN AMBERS

in the Victoria and Albert Museum

Marjorie Trusted

Victoria and Albert Museum
1985

ISBN 0 948107 13 8
Published by the Victoria and Albert Museum

Frontispiece 'The Judgement of Paris', *catalogue number 12*

Designed by Paul Sharp
Phototypeset in Bembo and
printed by Balding + Mansell Limited,
Wisbech, Cambridgeshire

Contents

Acknowledgements

In the course of preparing this book I have had great help, encouragement and advice from colleagues and friends. In particular I should like to thank all the members of the Departments of Metalwork and Sculpture at the Victoria and Albert Museum: Shirley Bury, who encouraged me to pursue the work, and Philippa Glanville, Marian Campbell and Anthony North, who made many helpful suggestions. Anthony Radcliffe, Malcolm Baker, Paul Williamson and Anna Somers Cocks painstakingly read through parts of the text, and gave me a great deal of invaluable constructive criticism. In addition, Anthony Radcliffe not only encouraged me during the course of the catalogue, but had initially suggested the idea, for which I am very grateful. Michael Snodin of the Department of Prints and Drawings helped me with ideas on engraved sources. John Larson, Carol Galvin, Graham Martin and Diana Heath aided me in the analysis of materials and construction of the objects. Stanley Eost photographed many of the pieces. I am grateful also to Christine Darby and Deborah Sinfield who patiently typed and re-typed the text. I should like to thank Hugh Tait at the British Museum, Godfrey Evans of the Royal Scottish Museum, Dr Mogens Bencard and Neils Jessen at the Rosenberg Collection, Copenhagen, the curatorial staff at the Grünes Gewölbe, Dresden, the Landesmuseum, Kassel, the Museum für Kunst und Gewerbe and the Altonaer Museum, Hamburg, Dr Johanna Lessmann of the Herzog Anton Ulrich-Museum, Brunswick, Dr Christian Theuerkauff of the Staatliche Preussischer Kulturbesitz, Berlin, Dr Lorenz Seelig of the Schatzkammer der Residenz, and Dr Peter Volk of the Bayerisches Nationalmuseum, Munich, Dr Helmut Trnek of the Kunsthistorisches Museum, Vienna, Hans Lepp and Ulla Lind of the Royal Collection Stockholm, Dr Helena Dahlbäck Lutteman, Dr Barbro Hovstadius and Lars Sjöberg of the National-museum Stockholm, Maj Nodermann of the Nordic Museum Stockholm, Professor Bertil Almgren of the Royal Scientific Society Uppsala, Dr Arne Losman and Johan Knutsson of Skokloster Castle. I am also indebted to two Polish colleagues, Janina Grabowska of Gdansk, and Elzbieta Mierzwinska of the Malbork Castle Museum. Barbara Dejlidko has translated articles in Polish, which has allowed me access to work I should not otherwise know. Antonia Boström has likewise translated Swedish pieces for which I am also grateful. Helen Fraquet has frequently discussed different types of amber with me, and her knowledge as a gemmologist has been very helpful. Claude Blair gave me some useful advice on cutlery. Alison and John Milbank's knowledge of Christian iconography was also invaluable. Generous grants from the British Antique Dealers Association and the British Council enabled me to undertake study-trips to Germany, Denmark and Poland to pursue my research. Finally I should like to thank John Parker for his constructive criticisms and encouragement throughout.

Foreword

This catalogue forms part of a series of publications devoted to the less well-known parts of the National Collection of Sculpture, housed in the Victoria and Albert Museum. Although it is generally recognised that the Museum has outstanding holdings of Italian sculpture and medieval ivory carvings (both of which already have catalogues devoted to them) there is a considerable amount of important material held by the Department which is still unpublished, and which consequently has not received the attention it deserves in general accounts of the History of Art. The aim of the series is to make these coherent groups more accessible to both the scholarly and general public and, as in the case of this catalogue, sometimes to offer a convenient introduction to the subject covered. Romanesque sculpture was catalogued in 1983 and the present volume will be followed by publications on the carvings of the Embriachi workshop in 15th-century Venice, Northern Gothic Sculpture 1200–1400, and German medals.

ANTHONY RADCLIFFE
Keeper of Sculpture

Fig 1: Illustrated title page from P. J. Hartman;
Succini Prussici Physica et Civilis Historia,
Frankfurt, 1677. Reproduced by kind permission of the
British Library.

Fig 2: The Baltic Coast in the 17th Century

Introduction

The Formation of Amber: Ideas and Myths

Amber has always been regarded as an exotic substance; from Prehistoric times onwards it was considered to possess magical healing powers. Amulets of amber have been found dating from 3500–1800 BC;[1] in Roman times Pliny the Elder remarked on its efficacy in healing throat infections.[2] The two earliest books on amber, written in the mid 16th century, were by doctors, explaining its medicinal qualities,[3] while a 17th-century English book recommended it as a test of virginity.[4] Because of its static electricity, when the mysterious forces of electricity were beginning to be investigated in the 17th century, the Greek word for amber, elektron, was chosen as the root of the English, electricity. Today in Mexico, amber beads are sold to ward off the influence of the 'evil eye'.[5]

Until recently the age and origins of amber were unknown.[6] One of the earliest myths on how it was formed is recounted in Ovid's story of Phaethon, son of Apollo the sun-god. On failing to control the horses of the chariot of the sun, he was struck by Jove's thunderbolt and fell to earth. His sorrowing sisters wept so profusely that they were metamorphosed into trees, and their tears into amber.[7] This myth is depicted on cat no 2. As the story implies, amber is a resin; it is actually a fossilised resin, and, depending on which part of the world it comes from, can be anything from 20 million to 120 million years old: Sicilian amber for example is about 25 million years old, while amber from New Jersey, U.S.A. is about 100 million years old.[8] However, European amber works of art are almost invariably made of Baltic amber, as are all the works in this catalogue. This dates from the Eocene period during the Tertiary era, and is about 40 million years old.[9] It was formed when what is now the Baltic Sea was land (the Baltic is only 100,000 years old), and dense forests of deciduous and coniferous trees grew there in a semi-tropical climate. For reasons unknown, the trees exuded abnormally large quantities of resin, and this in time fossilised to form amber. The climate cooled, and the trees, whose species even now are unknown, became extinct. With the coming of the Ice Age, some of the amber was carried by glaciers and rivers to what is now Poland and North East Germany; this can be mined from a stratum known as Blue Earth. Some remained on what became the Baltic Sea bed, and is dredged up in nets (see fig 1), or washed up on the beach by storms.

Although Aristotle, and later Pliny, had realised that it was a resin,[10] not until this century was the exact age of amber established.[11] The long geological periods involved in its formation explain why as late as 1734 Alexander Pope could write:

'Pretty! in Amber to observe the forms
Of hairs, or straws, or dirt, or grubs, or worms;
The things, we know, are neither rich nor rare,
But wonder how the Devil they got there?'[12]

Although none is present in the amber used for works of art, inclusions (the 'grubs or worms'), provide invaluable evidence today to paleontologists and entomologists studying forms of animal and plant life (some now extinct) which were trapped in the sticky resin.

The Working and Trading of Amber during the Medieval Period and the Rôle of the Teutonic Order

In Europe the history of amber carving dates back, as mentioned above, to Prehistoric amulets. Two amber cups excavated from Bronze Age sites in England are today to be seen in Dorchester and Brighton.[13] It was also treasured by early civilizations: beads were excavated at Mycenae,[14] Homer describes Menelaus' palace as decked with amber,[15] and in Roman times jewellery, figurines, reliefs and pots were carved.[16] Although some amber has been found from the Merovingian period and later,[17] the next important era of amber-working following Roman times is the medieval period, when guilds of *Paternostermacher* were set up in Bruges (by 1302) and Lübeck (before 1360), and later elsewhere, to manufacture amber paternosters or rosaries.

The rosary makers were supplied by amber from the Baltic coast, which from the late 13th to the late 15th century was controlled by the Teutonic Order, a charitable Catholic Order founded during the 12th century to fight in the Crusades. In 1211 they returned to Europe, a strong military force, and in 1225 their aid was requested by Conrad, Duke of Masovia to subdue rebellious Prussians along the Baltic coast. After their success, they conquered the territory of Prussia (see fig 2), and built their chief stronghold at Marienburg (Malbork) in 1309, although by 1457 the residence of the Grand Master was established further east at Königsberg (Kaliningrad).[18] Recognising amber as a valuable raw material, they vigorously controlled its supply. The inhabitants were paid for any amber found, but were ruthlessly punished if they kept, sold or worked it within the Knights' territory.[19] Most of the raw material was exported to Bruges and Lübeck where the Guilds used it to turn beads for rosaries. A few contemporary paintings illustrate what appear to be amber rosaries, although it is often difficult to tell whether it is amber, glass or even coral which is depicted. Probably the most famous is the Arnolfini Marriage portrait of 1434 by Jan van Eyck, in the National Gallery London, where an amber (possibly glass) rosary is seen hanging up on the wall.[20]

Amber was treasured too in medieval France and Burgundy: the inventories of Charles V show that in 1379 he owned a group (or relief) in amber of the Holy Family with the Three Kings and St. Anastasius, and an amber Madonna with a gold coronet garnished with pearls. In 1390 his inventory listed a statuette of St. John the Baptist, and in 1399 a knife with an amber handle. The Duchess of Touraine's inventory of 1389 listed a figure (or relief) of St. Margaret and the dragon mounted in silver-gilt, and three knives with handles of amber. The Duc de Berry owned in 1412 a relief of the Madonna mounted on wood studded with gold 'marcs', and in 1416 his inventory included a Madonna with a gold crown under a baldacchino. The Treasury of the Louvre housed a crucifix of amber in 1418. Charlotte of Savoy owned a small relief of Christ's head (a 'vera icon') in 1483. Charlotte d'Albret's inventory of 1514 listed a dozen amber paternosters, and in 1558 the inventory of Philip II included a bottle of amber mounted in silver gilt.[21] The Teutonic Order may have sold some of these or sent them as diplomatic gifts, as they employed artists to carve amber at their court in Königsberg.[22]

Some raw amber reached London at this time; rosary beads have been excavated on the site of Baynard's Castle dating from the early 14th century. These were discarded as refuse and are in various stages of working, from raw amber to polished, turned and drilled beads. The location of the workshops is not known for certain, but may have been in the vicinity of Paternoster Row, near St. Paul's, in the City of London.[23] Other amber beads dating from c. 1400 and from the 16th century have been found on the same site, and are now housed in the Museum of London.[24] Although some Baltic amber is washed up in small fragments on the East coast of England, this would not be enough to supply a workshop, thus amber must have been imported, probably via Bruges or Lübeck.

During the 15th century, the power of the Teutonic Order steadily declined, largely due to defeats suffered at the hands of the Polish army, and in 1525, the Grand Master, Albrecht of Hohenzollern, declared himself a convert to Lutheranism, disbanded the Prussian wing of the Order, and became Duke of Prussia, although the duchy was under Polish suzerainty.[25] As under the rule of the Teutonic Order, the Prussian inhabitants were forbidden to work amber.

Albrecht's conversion however was of great importance for the history of amber working; he is said to have called it 'Prussian silver', and largely due to his active encouragement, arts and culture, and the art of amber carving in particular, flourished, while his court at Königsberg fostered numerous skilled artists.[26] Protestantism meant that rosaries were less in demand, and more secular courtly objects such as tankards, goblets, handles for cutlery, candlesticks, caskets and gamesboards were made. Albrecht also lowered the cost of obtaining the raw material; previously the shore dwellers had been paid for amber with money and salt: he decreed that they should be paid only with salt. He also expanded the market for amber by selling it not only to Bruges and Lübeck, but to merchants in Königsberg, Danzig, Augsburg, Breslau and Antwerp. In 1533 an entrepreneur, Paul Jaski of Danzig was established, with the Duke's approval, as the sole distributor of amber to the merchants in other cities.[27] Amber was being traded even further afield with the Middle East: in the mid 16th century Armenians travelled to Königsberg and bartered silk carpets for raw amber.[28] An amber casket was taken as a diplomatic gift to Osman II of Turkey by a Polish Duke Krzysztof Zbaraski in 1622, [29] and Turkish mouth-pipes mounted in gilt enamels (probably dating from the late 18th or early 19th century) can be seen in the Malbork Castle Museum, late examples of objects almost certainly made for export to the Middle East.[30]

Amber Carving in Cities and Courts during the 17th Century

As amber working flourished, guilds sprang up along the Baltic coast.[31] Königsberg however was probably the leading centre for amber carving in the late 16th and early 17th century; until 1618 it was the seat of the Prussian Court as well as a thriving centre for goldsmiths.[32] The earliest ambers which can be precisely dated, a set of plates mounted in silver gilt, were made there for Georg Friedrich von Ansbach, the Regent of Prussia, probably by the goldsmith Andreas Knieffel and the amber-worker Stenzel Schmidt in 1585.[33] Most of the artists however remain anonymous or at best shadowy figures: Schmidt was evidently one of the most important of the late 16th century, but nothing is known of him beyond his name and activity at Königsberg. Georg Schreiber, (active c. 1617–1643), who was one of the founders of the Guild at Königsberg in 1641, signed several works, and others have been attributed to him on the basis of these. His finest piece is an altar in the Museo degli Argenti at the Palazzo Pitti, Florence, signed and dated 1618/19.[34] In our collection, cat no 6 may be influenced by his workshop. Other artists at Königsberg, such as Jacob Heise (active mid 17th century), provide a rough stylistic touchstone for other works (see entry for cat no. 7).

In Poland, at Danzig (Gdansk), about one hundred miles west of Königsberg, secular items were made for the Polish court, along with devotional objects such as statuettes of the Madonna and house altars, and contemporary sources indicate that in the 17th century, carved ambers were a highly regarded aspect of Polish art and craftsmanship.[35] An autonomous free city within Poland, Danzig flourished just as Königsberg seems to have declined somewhat in prestige as a centre for amber-working. In 1618, the duchy of Prussia was assimilated by the Elector of Brandenburg, whose court was held in Berlin. Königsberg however still thrived as a trading centre, and retained some ceremonial importance: the Great Elector made it his residence during the Thirty Years War, and Frederick III was crowned there as King Frederick I of Prussia in 1701.[36] The Guild at Danzig had been founded in 1477,[37] and it was this city, rather than Königsberg, which attracted most commissions from the Brandenburg court.

In 1677 the Great Elector commissioned from the Danzig artist Nicholas Turow a magnificent amber throne (of which only fragments and a drawing survive), for the Hapsburg Emperor Leopold I of Austria.[38] One of the artists who may have assisted on this project was Christoph Maucher (b. 1642; d. after 1701), who settled in Danzig in about 1670, and who himself secured further prestigious commissions from the Brandenburg Court.[39] The two fine works in the collection (cat nos 12 and 13) testify to his skill as a sculptor. He never however became a member of the Guild, nor even a citizen of Danzig, and the Guild blamed its decline in the early 18th century to his success in gaining commissions denied to its members. He was in effect a court artist who never resided at court. Amber was a great source of profit for the Elector of Brandenburg, who was selling it unworked for substantial sums during the 1690s. John Houghton, a contemporary Englishman, wrote in 1696: 'It is certainly a mystery to this day, whence all the amber comes, that is gathered yearly on the Prussian coast, which amounts to the

Elector's treasury (all expences deducted) about ten thousand pound sterling yearly.'[40]

Danzig must have continued to be highly regarded as a centre for amber-working, for in 1723 a French dictionary of commerce noted: 'C'est de Pologne et de Hongrie que vient l'Ambre le mieux travaillé & où aussi il se vend le plus cher'.[41] In 1709 two Danzig artists, Ernst Schacht and Gottfried Turow, were commissioned to embellish the amber room constructed for Frederick I in Berlin two years before by Gottfried Wolffram of Denmark.[42] This was the finest, yet perhaps the most decadent example of courtly commissions of amber, consisting of walls veneered with panels of amber, carved mirror-frames, reliefs and capitals. It became one of the grandest of diplomatic gifts too, when Frederick presented it to Peter the Great of Russia, to cement the alliance between Russia and Prussia against Sweden. By 1755 it had been erected in the Summer Palace at Zarskoje Sselo, but it was removed for safety to Königsberg in crates during the Second World War, and has subsequently been lost, perhaps destroyed at the end of the war.

Amber in England

Although most artists appear to have been German, and Königsberg and Danzig the leading centres, there is some evidence that amber was being worked in London too, after the medieval period, for there was certainly a market for ambers in London which was being supplied in the 17th and 18th century. By 1656 a duty had been fixed for amber, which implies that a substantial amount (either raw or worked) was already being imported,[43] and cutlery handles for blades made in London were being manufactured by 1638. It is uncertain whether they were worked in England or Germany, (see cat nos 33–36) although it is possible that there were hafters in London using amber at about this time.[44]

Houghton (mentioned above) bemoaned the fact that amber was not worked in England, although he may not have been aware of handles being made earlier in the century. He writes,

'As for (amber) beads, I do not hear that we make any in the kingdom, neither do we cabinets, to our shame be it spoken; I question whether there be one man in *England* knows how to work it (amber). I once had occasion for one, and could find none save Captain *Choke*, the inventer of necklaces for breeding teeth easy, (and now he is dead, I question if there be another).

'Methinks we should take off all duties from the simple *amber*, and give all encouragement to further its *manufacture*, which I persuade myself might be very considerable, if well managed.'[45]

Works of amber were certainly collected in England: the inventory of the collection of Anne, Viscountess of Dorchester, at Gosfield Hall, Essex, lists in 1638:

'I embroydred globe of glasses and an amber cupp in the middle of it . . . £10.0s.0d.'
'2 amber cupps, 2 candle sticks of amber, 1 amber spoone, 2 amber dishes and divers other odd things (in the Cabinet in the room next my Ladies Chamber) . . . £2.15s.0d.' and '1 amber cupp sett in silver and purple case of velvet . . . £2.0s.0d.'[46]

These were probably manufactured in Germany and imported, but the number of objects and the relatively high prices for which they were valued indicates that they were treasured items. Two other notable English collections which included amber were the 17th-century cabinet of curiosities owned by John Tradescant, and that of the Romantic collector William Beckford. The Tradescant collection, most of which is now housed in the Ashmolean Museum, has several items such as a necklace, rosaries and some small carvings.[47] In 1801 William Beckford owned at Fonthill Splendens an amber cabinet reputed to have been made for the Princess of Bavaria (or Queen of Bohemia) in 1665, which stood in the centre of the Library. Along with the rest of the contents of Fonthill Abbey, it was auctioned in 1823, and its present whereabouts are unknown.[48]

Two amber cabinets were at one time in the Royal Collection. A book on London and its environs published in 1762 refers to 'a curious amber cabinet, presented by the King of Prussia to queen Caroline' which was on display at Windsor Castle. This cabinet is mentioned again in *The English Connoisseur*, published in 1766.[49] Caroline of Ansbach (b. 1683; d. 1737) was the wife of George II (ruled 1727–1760); this presentation is one of numerous examples of diplomatic gifts of ambers made in the 17th and 18th century. The *English Connoisseur* continues, 'There is here likewise Queen Caroline's China Closet, filled with a great variety of curious china elegantly disposed . . . And in this Closet is also a fine amber cabinet, presented to Queen Anne, by Dr Robinson, Bishop of London, and plenipotentiary at the congress of Utrecht'. Neither cabinet is now in the Royal Collection, although a casket in the Museum, cat no 16, was donated by HRH Princess Louise in 1926, and it is interesting to speculate whether this was Queen Anne's or Queen Caroline's 'amber cabinet'. However, the proliferation of ivory reliefs on ours, not mentioned in the description of either of the cabinets at Windsor suggests that it is not.

A lace-making device, of the late 17th or early 18th century is inscribed 'Paul Morthorst, Londen (*sic*)' along with German mottoes; this may also have been made in Germany, (hence the spelling of London) but for an English customer.[50] However, by the mid 18th century several London dealers and craftsmen advertised on their trade-cards their abilities in manufacturing curiosities in amber.[51] The Guilds along the Baltic were in the meantime declining rapidly;[52] amber had become fashionable, and inevitably before long it went out of fashion. Few ambers of artistic value date from later than the mid 18th century: the amber head in the collection (cat no 21) is one of the last.

In late Victorian and Edwardian England amber jewellery was in vogue, and today in Poland it is used to make necklaces, bracelets and earrings, sometimes mounted in silver.[53] The Museum houses a silver bowl made in 1934–5 (not included in this catalogue) which has a finial set with amber.[54]

Techniques of Amber Working

Amber is of about the same hardness as ivory, but although it is easy to carve, great skill is required to prevent fractures caused by internal faults. Relief carving on panels, and sculpture in the round are two of the simpler and most popular techniques of working amber, and many examples can be seen in the collection, for instance reliefs on cat nos 6 and 7, and the sculpted figures cat nos 12 and 14. Some works illustrate the close links between amber and ivory: an amber cup of the late 17th century in the Kunsthistorisches Museum in Vienna is carved with a relief based on a painting by Rubens, as is an ivory cup by Artus I Quellin (1609–1668).[55] In the collection at Rosenborg, in Copenhagen are two identical pairs of putti, one in amber and one in ivory, carved by the court artist, Lorenz Spengler in the mid 18th century.[56] Although ivory and amber are sometimes juxtaposed in Königsberg works (cf cat nos 8, 25 and 26), it is a combination more usually evident in works from Danzig, particularly caskets, where it can be seen how the two materials' contrasting colours mutually enhance each other. The ivory reliefs are applied to wooden carcases which support the amber panels, and they are often pierced and set against a horn, painted foil or mica backing (cf cat nos 10, 11 and 16). This usage also accords with the fact that ivory and amber could be worked by the same artist, although whether this was always the case is not clear. On some works stylistic differences between the ivory and amber elements suggest a collaboration between an amber and an ivory worker, (see cat no 16). Other methods of working amber are peculiar to it, and exploit its colouring and translucency. When first mined or dredged up and cleaned it is a clear golden yellow, unless it contains impurities: these can give it colourings of brown, green, blue or even black. On being exposed to the air, the golden yellow slowly darkens over the years to a ruby red. Some amber pieces are cloudy and opaque, and some are of a dense white, resembling meerschaum. This last type, white amber, is created by microscopic droplets of water or air bubbles being caught up in the material.[57] It was much sought after for works of art, particularly the faces and hands of clear amber figures; (cf the figures on cat nos 9 and 11).

Six main decorative techniques were used:

1   Clear panels were set over small pierced reliefs of white amber, set in their turn against gold-coloured metal foil, horn, or mica. Sometimes these are on a wooden structure (as in cat nos 2 and 3); sometimes, as on cutlery handles, they are set in a double-sided cavity between two clear panels (see cat no 35). The relief is most usually white amber, although under the clear amber it is often difficult to distinguish from ivory or even meerschaum, particularly if the clear panels are suffering from surface crazing. In cases where the clear amber has been broken or lost however, the relief is almost invariably found to be white amber (although cat no 8 is an exception to this rule), and this is not surprising, for in Königsberg, where this technique was most often employed, white amber was available and highly valued.

2   Clear panels were cut in intaglio from behind. They were set so that, if the light shone on to them, the design almost had the appearance of Oriental lacquer

work, a combination of dark etched lines and gold. (See cat nos 8 and 17). The intaglio designs may be tendrils, arabesques or even mottoes or scenes.

3   The clear panels were gouged out from behind in a hemisphere; these (and sometimes the panels with intaglio designs also) could be placed over gold-coloured metal foil set on a wooden carcase. The play of light heightens the decorative design, and the reflective quality gives an ambiguity to the design or hemisphere, so that the panel seems to have a luminous depth. (See cat nos 8, 10, 11 and 18.)

4   The clear panels were placed over foil painted with a design or the name of the artist, owner, or date (see cat nos 2, 3 and 25). The foil is painted in gold, black, red and green, colours which vividly shine through the amber.

5   A similar technique, in which the underside of the amber panel is painted and placed over foil has been compared to *verre églomisé*, and has been called églomisé in books on amber. It is similar to Boulle work, where painted tortoiseshell is placed over foil.[58] Like the first technique described, this is most commonly seen in works from Königsberg.

6   In Danzig, one further technique was developed which enhanced the contrasts between opaque cloudy coloured panels and clear ones (see cat nos 9, 11). Occasionally a mosaic relief or picture was formed of different coloured panels (see cat no 9, see also the Italian example, cat no 22). The panels were mounted on wood, a technique known as incrustation, which was especially popular in the late 17th century. Although this is seen in the gamesboards from Königsberg of the late 16th and early 17th century, in Danzig the wooden carcase was used to construct monumental works (such as the throne for Emperor Leopold I mentioned above) from quite small pieces of amber. Most amber is found in relatively small fragments, rarely larger than four centimetres across, and artists sometimes had to wait for pieces large enough for a particular work; on the back of a mirror in Burnswick, the artist Johann Köster (or Küster) has written that he waited five years for the amber necessary![59]

Sources for Works in Amber

Few designs for works in amber have survived; the little evidence there is suggests that the amber workers themselves usually made the drawings, although they were often inspired by works in other media.[60] As mentioned above, ivories may have provided parallels for sculptural works, and several artists, such as Christoph Maucher, Wilhelm Krüger, Jacob Dobbermann, and Lorenz Spengler, were both ivory and amber carvers. The forms of the house-altars reflect sculpted stone monuments (see cat nos 9 and 10) and this emulation of larger-scale and sculptural forms seems to be particularly associated with Danzig.

The gamesboards (see cat nos 2 and 3) reflect contemporary wooden gamesboards inlaid with ivory, mother o' pearl, and silver; amber was evidently considered an exotic embellishment, and it was in Königsberg particularly that the gamesboards were made. There too, ambers reflected contemporary goldsmiths' work, such as the tankards and cups and covers made in the late 16th and early 17th century.[61] Many amber examples of these are actually mounted in gilt metals, and are evidence of collaboration between goldsmiths and amber-workers; (see cat nos 6 and 7, and the plates mentioned above of 1585 by Stenzel Schmidt and Andreas Knieffel.)

The techniques of inlay, incrustation and dowelling recall carpenters' and joiners' work, and amber caskets and cabinets, such as cat nos 17 and 18, are reminiscent of tiered wooden caskets.[62]

Ceramics also provided inspiration, for in the early 18th century, several ambers were carved of figures reclining in shell-shaped baths, which imitate the work of the 16th-century French potter, Bernard Palissy.[63]

Few of the engraved sources from which many amber reliefs and ivory ones set on amber caskets and altars may depend have been identified, although German and Flemish 16th- and 17th-century engravings seem to have been widely used (cf cat nos 9 and 16).

The Kunstkammer

These various influences and sources of inspiration for ambers, and the interaction with styles and methods used in other media may have been accentuated by the intellectual and artistic atmosphere of the *Kunstkammer*. These cabinets of curiosities owned by rulers and wealthy burghers flourished from the 16th to the early 18th century, coinciding with the great period of amber-working.[64] They consisted of small highly-worked objects such as bronzes, ivories, silver-gilt cups and covers, and exotic items such as tropical birds' feathers, ostrich-eggs and unicorn's horns (the latter invariably a narwhal's tooth). Carved ambers fulfilled both the essential demands of the *Kunstkammer* object, being not only highly worked but made of a mysterious substance, so they were particularly treasured pieces. Like nautilus shells, coconut shells, rhinoceros horn and ostrich eggs, amber was thought to have the power to indicate if a food or drink was contaminated with poison, and this may explain why so many cups, goblets, and tankards were made from it in the early 17th century[65] (see cat nos 6 and 7). Frequently artists were appointed by the court to make objects for the collection, such as Jacob Dobbermann at Kassel, Wilhelm Krüger at Dresden, and Lorenz Spengler at Copenhagen, and these artists often worked in more than one medium.[66] The interplay of ideas would naturally encourage stylistic influences between ambers and ivories, goldsmiths' work, ceramics and so on. Engraved sources used in one art might become available to those practising another. The collections today in Vienna, Munich, Dresden, Kassel and Copenhagen bear witness to the extraordinary achievements of these artists and their patrons.

The Collection

The collection of ambers in the Victoria and Albert Museum is unusual for its range and variety. It includes caskets, figurative reliefs, small-scale sculpture, altars, carved vessels, a rosary, and cutlery handles. The period covered spans the 15th to the mid 18th century, and there are objects worked in Italy, The Netherlands and England as well as Germany. Because of the way the collection was formed, by purchase, gift or bequest from the mid 19th century onwards, the provenance of the objects is highly varied. This contrasts with the great Continental collections of ambers, which descend from the 16th- and 17th-century royal *Kunstkammern* mentioned above.[67] Although there are many donors, the most generous was Dr W.L. Hildburgh FSA; his gifts are among the finest in the collecton.

Amber is a fragile material, and needs to be kept under dark, stable conditions in a fairly cool atmosphere of a constant humidity (approximately 55 rh), away from dust; many works deteriorate because they are exposed to light and heat. An airtight museum case in a dimly lit, cool gallery is ideal, although even so they have to be kept under constant observation. Many in the collection have undergone restoration, while some of the inscriptions and reliefs have become illegible.

The catalogue entries cover only the European works in the collection, although a handlist of the ambers in the Far Eastern Department is included in an appendix compiled by Craig Clunas.

Problems of Attribution

As no authentically signed or documented ambers are housed in the Victoria and Albert Museum, attributions to particular German cities rely on analogies and circumstantial evidence. Some of the stylistic parallels may themselves depend on further analogies for their identification. The basis of knowledge is provided by the few signed works or drawings by artists known to be active in Königsberg or Danzig. These show two distinct styles, the elements of which are discussed in the sections on techniques and sources. In general, the works from Königsberg are more closely modelled on goldsmiths' work, while those from Danzig tend to be more sculptural and monumental. The catalogue entries give more detailed information on the reasons for particular attributions.

Literature

Innumerable books and articles have been published on the scientific aspects of amber: the different types, the inclusions found in it, and what these tell us about early geological eras.[68] Fewer have been written on the art historical side: Otto Pelka's pioneering works on the Guilds (in 1917) and on amber works of art (in 1920) initiated serious study of the subject, but it was Alfred Rohde's seminal work of 1937 which established important ideas about the development of amber-working and the artists who carved it. The present catalogue is greatly indebted to the work of these two scholars. The interest in amber in Germany in the 1920s and 1930s was however perhaps not wholly innocent; it symbolised German (especially Prussian) nationalism, evinced in the subtitle of Rohde's book, 'ein Deutscher Werkstoff' (a German raw material). One work, published in 1937, even managed to introduce the topic of the Versailles Treaty into its opening![69] Gisela Reineking von Bock's survey of 1981 has excellent illustrations, but its text relies heavily on Rohde, and also, perhaps again, overemphasises the importance of Prussia. The most recent work is Janina Grabowska's study of Polish ambers of 1982, which somewhat corrects the balance by highlighting the activity of Danzig, and brings to the fore a great deal of important new material based on documentary evidence, although unfortunately the sources are often impossible to follow up. Some of the books published in English (such as G.C. Williamson's publication of 1932), are good introductory works, although they lack the strict scholarship of Pelka and Rohde. P. Rice's book published in 1980 is a superficial survey of the scientific and artistic aspects of amber.

Many important articles have been published in the last twenty years or so, which have broadened the field of study considerably, and established a few exact dates or facts (for example the publications of Kosegarten, Aschengreen-Piacenti and Baer). These have often elucidated problems of identification associated with ambers in this collection.

Notes

1 Reineking von Bock, fig 22. Examples are also illustrated in Grabowska.

2 Pliny, *Natural History*, XXXVII, 12.

3 A. Aurifaber, *Succini Historia*, Königsberg, 1551, and S. Göbel, *History und eigendtlicher bericht von herkommen Vrsprung und vielfeltigen brauch des Börnsteins neben andern saubern Berckhartzen so der Gattung etc. aus guten grundt der Philosophy*, 1566.
   These are referred to in Rohde, pp 17, 18; the present author has been unable to obtain copies of either.

4 'White amber; one kind thereof (thrown by the floating sea on the Pruthian shore) which being given to drink in wine unto a fasting wench, will force her to piss, if she have lost her maidenhead'. R. Cotgrave, *A French and English Dictionary*, London, 1673.

5 D. Schlee, *Bernstein-Raritäten: Farben Strukturen Fossilien Handwerk*, Stuttgart, 1980, p 71.

6 See C.S. Beck, M. Gerving and E. Wilbur, 'The Provenience of Archaeological Amber Artifacts' Parts I and II in *Art and Archaeology Technical Abstracts*, VI, 1966 and 1967, nos 2 and 3, pp 215–302, and 203–273 for a bibliography on the complex history of ideas about the nature of amber.

7 Ovid, *Metamorphoses*, II.

8 Schlee, *op cit*, pp 9–10. Amber can also be found in Rumania, Burma, the Dominican Republic and elsewhere.

9 See S.G. Larsson, *Baltic Amber – A Palaeobiological Study*, Klampenborg, 1978, pp 26 ff.

10 Aristotle, *Meteorologika*, IV, 10: 'the animals enclosed in (amber) show that it is formed by solidification'. Pliny, *Natural History*, XXXVII, 12.

11 As late as 1932, G.C. Williamson could write: 'Amber is many thousands of years old, was formed perhaps a million years ago'. Williamson, p 15.

12 A. Pope, *Epistle to Dr Arbuthnot* (written 1731–4, published 1735, lines 169–172).

13 The Dorset County Museum houses one, now in a fragmentary state, dug up from the Clandown Barrow. I am very grateful to R.N.R. Peers for this information.
   The other is in the Brighton Art Gallery and Museum. See E. Curwen and E.C. Curwen, 'The Hove Tumulus', *Brighton and Hove Archaeologist*, II, 1924, pp 20–28 and Plates V, and VI. I am very grateful to J.I. Hadfield for this information. See also S. Piggott (Ed), *The Dawn of Civilization*, London, 1961, p 353.

14 A.J.B. Wace, *Mycenae, an Archaeological History and Guide*, Princeton, 1949, p 108.

15 Homer, *Odyssey*, IV, 53–84. Some ambiguity remains, as elektron can mean an alloy of silver and gold, but here it is thought that amber is intended.

16 See D.E. Strong, *Catalogue of the Carved Amber in the Department of Greek and Roman Antiquities*, London, 1966. See also Pliny, *Natural History*, XXXVII, 12, where he says that an amber statuette is worth more than several able-bodied slaves.

17 Eg a necklace in the Römisch-Germanisches Museum, Cologne, inv 134. See Reineking von Bock, fig 54. A few

amber objects of the 11th and 12th century have been
excavated in Lund in Sweden. See A.W. Martensson, *St Stefan
i Lund*, Lund, 1981, figs 107–8.

18 See Rohde, *Königsberg*, p 22.

19 See Rohde, pp 13–16. See also *Encyclopaedia Britannica*,
Chicago, 1977, IX, p 913, and A.R. Chodynski, *Malbork*,
Warsaw, 1982, p. 99.

20 See M.J. Friedländer, *Early Netherlandish Painting*, Leyden,
1967, plates 20, 21.

21 See V. Gay, *Glossaire Archéologique du Moyen Age et de la
Renaissance*, Paris, 1887, I, p 28. See also Rohde, p 16.

22 Rohde, pp 14–15.

23 See V.K. Mead, 'Evidence for the Manufacture of Amber
Beads in London in 14–15th Century', *Transactions of the
London & Middlesex Archaeological Society*, XXVIII, 1977,
pp 211–214.

24 I am very grateful to Peter Marsden of the Museum of London
for allowing me to look at these beads, and to Philippa
Glanville for drawing my attention to them.

25 The King of Poland had been patron of the Order since 1477.
See Pelka, p. 42.

26 Rohde, pp 17 ff, and Rohde, *Königsberg*, p 2. See also
A.G. Dickens, *Reformation and Society*, London, 1966, p 74.

27 Rohde, p 17.

28 Rohde, p 13, Pelka, p 43.

29 Grabowska, p 20.

30 Illustrated in Grabowska.

31 Pelka, *Meister*, pp 12–32. Records do not survive for the exact
dates of the foundation of all the guilds. The Pomeranian ones,
Stolp, Kolberg, and Köslin, were founded in the 15th (Stolp),
and 16th century (Kolberg and Köslin). Danzig was established
in 1477, Elbing by 1539, and Königsberg in 1641. As
Chodynski notes, these guilds of *Paternostermacher* probably
manufactured rosaries in other materials as well as amber, and
the number of artists working in amber cannot be gauged from
the number of guild-members. Conversely, many artists who
were not guild-members worked in amber (see below). See
A.R. Chodynski, 'Spis Bursztynnikow Gdanskich od XVI do
wieku', *Rocznik Gdanski*, XLI, 1981, pp 193 ff.

32 See A. Rohde and U. Stöver, *Goldschmiedekunst in Königsberg*,
Stuttgart, 1959, *passim*.

33 These are now in the Rosenborg Collection, Copenhagen.
Rohde, pp 22–23 and fig 10. Rohde and Stöver, *op cit* fig 29.

34 K. Aschengreen-Piacenti, 'Due Altari in Ambra al Museo
Degli Argenti'. *Bollettino d'Arte*, III–IV, July–December, 1966,
pp 163 ff.

35 C.f. the Madonna given by the Abbot of the Oliva Monastery
near Danzig to the shrine of Jasna Gora in 1604, and the
gamesboard now in the Wawel Crown Treasury, Cracow,
given by a courtier to King Sigismund III Vasa in 1608. Many
other ambers made at Danzig have been lost or destroyed from
the Thirty Years War in the 17th century up to the Second
World War. A young Pole from Cracow writing from Italy to
his father in 1640 requested a work of amber as an example of
fine Polish art to give to a friend in Rome. See Grabowska, pp
18 ff and p 24. In the late 16th and first half of the 17th century
forty amber workshops were active in Danzig. See Chodynski,
Rocznik, op cit p 198, after M. Bogucka.

36 Anna, the daughter of Duke Albrecht's son, Albrecht Friedrich
(1553–1618), married Johann Sigismund, Elector of
Brandenburg (1572–1619) thus uniting the two territories into
what became the Kingdom of Prussia. See W.K. Isenburg,
Stammtafeln zur Geschichte der Europeänische Staaten, I–II,
Marburg, 1953, Tafeln 61, 62. See Rohde, *Königsberg*, pp 92,
103–104. The city evidently retained some importance as a
centre for amber-working, as in c. 1725 an amber cabinet was
constructed there to be given to the Elector of Saxony by
Frederick William I. See Rohde, pp 66–67.

37 Pelka, *Meister*, p 20.

38 W. Baer, 'Ein Bernsteinstuhl für Kaiser Leopold I. Ein
Geschenk des Kurfürsten Friedrich Wilhelm von
Brandenburg'. *Jahrbuch der Kunsthistorischen Sammlungen in
Wien*, LXXVIII, (N.F. XLII), 1982, pp 91–138.

39 See entries for cat nos 12 and 13 for detailed references.

40 J. Houghton, *A Collection for the Improvement of Husbandry and
Trade*, revised by R. Bradley, London, 1727, II, CCV, Friday,
3 July 1696, p 65.

41 J. Savary des Bruslons, *Dictionnaire Universel de Commerce*, I,
Paris, 1723, p 86. No Hungarian ambers are known, and
reference to them may be a mistake on the part of the
compiler.

42 See A. Rohde, 'Das Bersteinzimmer Friedrichs I in
Königsberger Schloss', *Pantheon*, XXIX, 1942, pp 200–203.
See also Pelka, pp 48–50, and pp 62–67.

43 See G. Foxcroft *et al*, *Book of Rates*, London 1656, p 1. I am
grateful to Graham Smith of H.M. Customs and Excise
Library Services for verifying this for me. Houghton implies
that most imported amber is unworked, but is sent 'out again
unwrought'. See Houghton, *op cit*, p 66.

44 The London Cutlers' Company records include an entry on the
apprenticeship of James Maes to a cutler on 17 June 1628, but
who was 'turned over' to Conrad Peeters, a jeweller and stone
cutter, on the same day. Peeters may have used amber to make
knife-handles. Maes became a Freeman of the Cutlers'
Company on 7 July 1635. I am grateful to G.I. Mungeam for
this information.

45 Houghton, *op cit*, p 66.

46 *Notes & Queries*, April, November, December 1953, pp 156,
470 and 517. See P. Thornton, *Seventeenth Century Interior
Decoration in England, France & Holland*, New Haven, 1978,
p 247.

47 See A. MacGregor (Ed.), *Tradescant's Rarities. Essays on the
Foundation of the Ashmolean Museum 1683 with a catalogue of the
Surviving Early Collections*, Oxford, 1983, pp 234 ff.

48 See *The Beauties of Wiltshire displayed in Statistical, Historical,
and Descriptive Sketches Interspersed with Anecdotes of the Arts*, I
London, 1801, pp 212–213, and C. Wainwright, 'William
Beckford's Furniture', *Connoisseur* CXCI, no 770, 1976, fig 10,

after J. Britton, *Illustrations Graphic and Literary of Fonthill Abbey Wiltshire*, 1823, and the sale catalogue for the auction: *The Unique and Splendid Effects of Fonthill Abbey*, Phillips, 23 September 1823, p 182, cat no 1041. I am grateful to Clive Wainwright for supplying this information.

49 A review of *London and its Environs described* . . . , in *The Critical Review: or Annals of Literature*, X, London, 1762, p 306. I am grateful to Malcolm Baker for drawing my attention to this.

50 This is now in the Malbork Castle Museum.

51 See A. Heal, *The London Goldsmiths 1200–1800. A record of the Names and Addresses of the Craftsmen, their Shop-Signs and Trade-Cards*, Cambridge, 1935, plates XV, XVIII, XLVI, LX and ·LXXIV.

52 The Guild at Lübeck continued until 1842, the Königsberg Guild declined from 1811 onwards, the Pommeranian Guilds continued until about the end of the 18th century, and the Danzig Guild until the mid 18th century. See Pelka, *Meister*, pp 8, 13–19, 20, 24.

53 See the illustrations in Grabowska.

54 Inv Circ 181 and a-1952. I am grateful to Jane Stancliffe for drawing my attention to this object.

55 The amber is in the Kunsthistorisches Museum, Vienna, inv 3542, and is unpublished. For the ivory, see *La Sculpture au Siècle de Rubens*, Musée d'Art Ancien, Brussels, 1977 (exhibition catalogue), p 155, cat no 117. The mounts are dated 1686, although the ivory is earlier. The composition is influenced by Rubens.

56 They depict Prudence and Fortitude. The amber ones are unpublished, but the ivory ones, which were part of a mounted medallion to commemorate the birthday of Frederik II of Denmark, 31 March 1759, are illustrated in P. Gouk, 'The Union of Arts and Sciences in the Eighteenth Century: Lorenz Spengler (1720–1807), Artistic Turner and Natural Scientist', *Annals of Science*, 40, (1983), p 423, fig 3.

57 See Larsson, *op cit*, pp 9–10.

58 I am grateful to Peter Thornton for pointing out this parallel

59 See Rohde, p 54.

60 The designs made by Michael Redlin of Danzig in 1688 are illustrated in Rohde, figs 23–28, and in Grabowska. The drawing of the throne by Nicholas Turow of Danzig in 1677 is in Baer, fig 57. Unpublished drawings of the mid 18th century by Marcus Tuscher for works by Lorenz Spengler are housed in Copenhagen.

61 See Rohde, figs 95–108, Reineking von Bock, figs 107–108, 111, 113–121 and cf Rohde and Stöver, *op cit*, figs 20–22, 24, 30–36.

62 Cf the late 17th-century wooden casket made for the *Paternostermacher* in Lübeck. See M. Hasse, *Lübeck Sankt Annen-Museum: Bilder und Hausgerät (Lübecker Museumsführer II)*. Lubeck, 1969, p 226, cat no 506, (illustrated on p 225).

63 Ambers usually attributed to Johann Christoph or Johann Kaspar Labhard (d. 1742 and 1726 respectively), in the Landesmuseum, Kassel, the Museum für Kunst und Gewerbe, Hamburg and the Rosenborg Collection, Copenhagen imitate Palissy's work. See Pelka, fig 83, Rohde, figs 240–247 and Reineking von Bock, figs 154–158. Cf G. de Rothschild and S. Grandjean, *Bernard Palissy et son Ecole*, Paris 1952, cat nos XXXVII and XXXVIII, Plate 36.

64 See J. von Schlosser, *Die Kunst- und Wunderkammern der Spätrenaissance. Ein Beitrag zur Geschichte des Sammelwesens*, Leipzig, 1908. See also B.J. Balsiger, *The Kunst- und Wunderkammern. A Catalogue Raisonné of Collecting in Germany France and England 1565–1750*, University of Pittsburgh Ph.D. 1970.

65 See S. Schade, 'Gefässe gegen Gift', *Informationen*, No 11, *Staatliche Kunstsammlungen Kassel*, November 1982.

66 See Pelka, *Meister*, pp 32–35.

67 Eg those at the Kunsthistorisches Museum, Vienna, the Museo degli Argenti, Palazzo Pitti, Florence, the Schatzkammer der Residenz, Munich, the Landesmuseum, Kassel, the Grünes Gewölbe, Dresden and the Rosenborg collection, Copenhagen. Other important collections are in Sweden: at the Royal Scientific Society, Uppsala, Skokloster Castle, and the Royal Collection, Stockholm.

68 See eg the bibliography in D. Schlee and W. Glöckner, 'Bernstein', *Stuttgarter Beiträge zur Naturkunde*, serie C, no 8, 1978, and the bibliography in Reineking von Bock.

69 K. Andrée, *Der Bernstein und seine Bedeutung in Natur- und Geisteswissenschaften, Kunst und Kunstgewerbe, Technik, Industrie und Handel*, Königsberg, 1937, pp 7–8.

Select Bibliography   *See also the list of Abbreviations of Works Cited in the Text. Other books and articles of incidental interest to the study of amber are cited in full in the footnotes.*

K. Andrée, *Der Bernstein und Seine Bedeutung in Natur- und Geisteswissenschaften Kunst und Kunstgewerbe Technik Industrie und Handel*, Königsberg, 1937.

K. Aschengreen Piacenti, 'Due Altari in Ambra al Museo degli Argenti', *Bollettino d'Arte*, III–IV, July-December, 1966, pp 163–166.

Ibid (Ed), *Il Museo degli Argenti a Firenze*, Milan, 1968.

K. Aschengreen Piacenti, H. Honour and others, *Ambre, Avori, Lacce, Cere, Medaglie e Monete*, Milan, 1981.

W. Baer, 'Ein Bernsteinstuhl für Kaiser Leopold I. Ein Geschenk des Kurfürsten Friedrich Wilhelm von Brandenburg', *Jahrbuch der Kunsthistorischen Sammlungen in Wien*, LXXVIII, (N.F. XLII), 1982, pp 91–138.

C.W. Beck, 'Authentication and Conservation of Amber: Conflict of Interests'. *Science and Technology in the Service of Conservation. Preprints of the Contributions to the Washington Congress*, 3–9 September 1982, pp 104–107.

C.W. Beck, M. Gerving and E. Wilbur, 'The Provenience of Archaeological Amber Artifacts', Parts I and II, *Art and Archaeology Technical Abstracts*, VI, 1966 and 1967, nos 2 and 3, pp 215–302 and 203–273.

A.R. Chodynski, 'Spis Bursztynnikow Gdanskich od XVI do Poczatku XIX Wieku', *Rocznik Gdanski*, XLI, 1981, pp 193–214.

Ibid, Malbork, Warsaw, 1982

A. de Foelkersam, 'L'Ambre Jaune et son Application aux Arts' (in Russian), *Staruie Ghodui*, XI, 1912.

J. Grabowska, *Amber in Polish History*, Edinburgh, 1978.

Ibid, Polnischer Bernstein, Warsaw, 1982 (also available in English).

J.G. Haddow, *Amber. All About It*, Liverpool, 1892.

P.J. Hartmann, *Succini Prussici Physica et Civilis Historia*, Frankfurt, 1677.

W. Hildburgh, 'An Amber and Ivory Altar', *Apollo*, XXX, 1939, pp 208–213.

J. Hildebrand and C. Theuerkauff, (Eds) *Die Brandenburgisch-Preussische Kunstkammer Eine Auswahl aus den alten Beständen*, Berlin, 1981.

A. Kosegarten, 'Eine Kleinplastik aus Bernstein von François du Quesnoy', *Pantheon*, XXI, 1963, pp 101–108.

B. Kosmowska-Ceranowicz and T. Pietrzak, *Znalezicka I Dawne Kopalnie Bursztynu W Polsce*, Warsaw, 1982.

B. Kosmowska-Ceranowicz and others, *Ambra Oro del Nord*, Ducal Palace, Venice, 1978 (exhibition catalogue).

S.G. Larsson, *Baltic Amber – A Palaeobiological Study*, Klampenborg, 1978.

V.K. Mead, 'Evidence for the Manufacture of Amber Beads in London in 14–15th century', *Transactions of the London & Middlesex Archaeological Society*, XXVIII, 1977, pp 211–214.

W. Meinhold, *The Amber Witch*, translated by Lady Duff Gordon, London, 1846.

M. Meinz, 'Die Bernsteinsammlung im Altonaer Museum', *Altonaer Museum in Hamburg Jahrbuch*, VIII, 1970, pp 9–38.

Ibid 'Ein nordostdeutscher Hausaltar mit Bernsteininkrustationen', *Altonaer Museum in Hamburg Jahrbuch* II, 1944, pp 143–160.

J.M. de Navarro, 'Prehistoric Routes between Northern Europe and Italy defined by the Amber Trade, *The Geographical Journal*, LXVI, no 6, December, 1925, pp 481–507.

O. Pelka, 'Die Meister der Bernsteinkunst', Nuremberg, 1917.

Ibid, Bernstein, Berlin, 1920.

Ibid, 'Zum Werk des Bernsteinmeisters Georg Schreiber in Königsberg', *Pantheon*, XVII, 1936, pp 27–29.

Plinius Secundus, Caius, *Natural History, with an English translation by H. Rackham*, X, London, 1962. Book XXXVII.

G. Reineking von Bock, *Bernstein. Das Gold der Ostsee*, Munich, 1981.

P.C. Rice, *Amber the Golden Gem of the Ages*, New York, 1980.

P.G. Rzaczynski, *Historia Naturalis Curiosa Regni Poloniae*, Sandomira, 1721.

Ibid, Auctarium Historiae Naturalis Curiosae Regni Poloniae, Danzig, 1735 (?).

A. Rohde, *Königsberg Pr.*, Leipzig, 1929.

Ibid, Bernstein, ein Deutscher Werkstoff. Seine Künstlerische Verarbeitung vom Mittelalter bis zum 18. Jahrhundert, Berlin, 1937.

Ibid, 'Das Bernsteinzimmer Friedrichs I in Königsberger Schloss', *Pantheon*, XXIX, 1942, pp 200–203.

A. Rohde and U. Stöver, *Goldschmiedekunst in Königsberg*, Stuttgart, 1959.

D. Schlee, *Bernstein-Raritäten: Farben Strukturen Fossilien Handwerk*, Stuttgart, 1980.

D.E. Strong, *Catalogue of the Carved Amber in the Department of Greek and Roman Antiquities*, London, 1966.

J. Svennung, 'De Latinska inskrifterna på K. Vetenskapssocietetens bärnstensbrädspel', *Kungl. Vetenskapssocietetens Arsbok, Uppsala*, 1960, pp 123–138.

C. Theuerkauff, 'Kaiser Leopold im Triumph wider die Türken . . . Ein Denkmal in Elfenbein von Christoph Maucher, Danzig', *Hamburger mittel- und ostdeutsche Forschungen*, IV, Hamburg, 1963, pp 60–93.

H. Thoma and H. Brunner (Eds), *Schatzkammer der Residenz München Katalog*, Munich, 1964.

M. Trusted, 'Four Amber Statuettes by Christoph Maucher', *Pantheon*, XLII, 1984, pp 245–250.

Ibid, 'Carved Ambers – Baltic Gold: Prussian Silver', *Victoria and Albert Museum Album* III, 1984.

Ibid 'Smart Lethieullier's Amber Tankard', *Apollo*, CXXI, no. 279, May 1985.

R. Verres, 'Der Elfenbein und Bernsteinschnitzer Christoph Maucher', *Pantheon*, XII, 1933, pp 244 ff.

S. Wallin, 'Skattkammargods', *Fataburen Nordiska Museets Och Skansens Arsbok*, 1945, pp 117–124.

Ibid, 'Bärnstenskonst hos Kungl. Vetenskapssocieteten i Uppsala', *Kungl. Vetenskapssocietetens Arsbok*, Uppsala, 1960, pp 71–121.

E. Werwein, 'Die Restaurierung eines Bernsteinkabinetts aus dem Späten 17. Jahrhundert', *Arbeitsblätter für Restauratoren*, Jg. XV, 1982, I.

G.C. Williamson, *The Book of Amber*, London, 1932.

Abbreviations of Works Cited in the Text

Baer   W. Baer, 'Ein Bernsteinstuhl für Kaiser Leopold I.
                              Ein Geschenk des Kürfursten Friedrich Wilhelm von
                              Brandenburg', *Jahrbuch der Kunsthistorischen Sammlungen in
                              Wien*, LXXVIII, (N.F.XLII), 1982, pp 91–138.

Bailey   C.T.P. Bailey, *Knives and Forks*, London, 1927.

Bartsch   W.L. Strauss (Ed.), *The Illustrated Bartsch*, New York,
                              1978–83.

Beard   C.R. Beard, 'Wedding Knives', *Connoisseur*, LXXXV,
                              1930, pp 91–97.

Blair   C. Blair, *The James A. De Rothschild Collection at Waddesdon
                              Manor: Arms, Armour and Base-Metalwork*, Fribourg, 1974.

Franke   I.O'Dell-Franke, *Kupferstiche und Radierungen aus der
                              Werkstatt des Virgil Solis*, Wiesbaden, 1977.

Grabowska   J. Grabowska, *Polnischer Bernstein*, Warsaw, 1982.

Hayward   J.F. Hayward, 'The Howard E. Smith Collection of
                              Cutlery', *Connoisseur*, CXXXIV, 1954, pp 164–173.

Hollstein   F.W.H. Hollstein, *Dutch and Flemish Etchings, Engravings
                              and Woodcuts ca. 1450–1700*, Amsterdam, 1949–1983.

Hughes   G.B. Hughes, 'Old English Wedding Knives', *Country Life*,
                              CV, 1949, pp 666–667.

Masterpieces   *Masterpieces of Cutlery and the Art of Eating*, Victoria and
                              Albert Museum, London, 1979 (Exhibition catalogue).

Pelka   O. Pelka, *Bernstein*, Berlin, 1920.

Pelka *Meister*   O. Pelka, *Die Meister der Bernsteinkunst*, Nuremberg, 1917.

Reineking von Bock   G. Reineking von Bock, *Bernstein Das Gold der Ostsee*,
                              Munich, 1981.

Rohde   A. Rohde, *Bernstein. Ein Deutscher Werkstoff seine
                              künstlerische Verarbeitung vom Mittelalter bis zum 18.
                              Jahrhundert*, Berlin, 1937.

Rohde, *Königsberg*   A. Rohde, *Königsberg Pr.*, Leipzig, 1929.

Trusted, *Album*   M. Trusted, 'Carved Ambers: Baltic Gold – Prussian
                              Silver', *Victoria and Albert Museum Album*, III, 1984.

Trusted, *Pantheon*   M. Trusted, 'Four Amber Statuettes by Christoph
                              Maucher', *Pantheon*, XLII, 1984, pp 245–250.

Welch   C. Welch, *History of the Cutlers' Company of London*,
                              London, 1916.

Williamson   G.C. Williamson, *The Book of Amber*, London, 1932.

Detail from Altar, catalogue no 9
The Adoration of the Shepherds

The Ambers

1   Pair of Cruets for The Mass

Lübeck or Bruges. Mid 15th to early 16th century.
Amber mounted in silver-gilt. Champlevé enamel coat of arms at the base of one cruet.
Height 12 cm Diameter of base 5 cm
4260 and 4261-1857. Department of Metalwork.
Bought in 1857. (Provenance unknown).

Surface scratches occur on the amber of both cruets, and the coat of arms at the base of one (4260-1857) is missing; otherwise they are in good condition.

These cruets are probably the earliest ambers in the collection, and the simplicity of their form reflects this. Each one is of turned opaque amber, which is bare of any decorative features itself, but is mounted in silver-gilt and adorned with a coat of arms, (one of these is now missing), of champlevé enamel.

The coat of arms has not been identified; it is blazoned gules, a lion rampant or, impaling argent a hunting horn proper. As an impaled shield usually represented a marriage (the wife's arms on the sinister), the wife's family could be Horn, the arms being a pun on the name. The crest is a crozier, which does not necessarily mean that the arms are those of a bishop, but implies that a bishop had been a member of the family.[1] The sacramental nature of the cruets is indicated by the engraving of 'A' for *aqua* and 'V' for *vinum* on the lid of each.

The form of the cruets suggests an approximate date by comparison with 15th- and early 16th-century goldsmith's work, and this accords with the way the amber has been worked.[2] Although a few records survive of carved 14th- and 15th-century ambers,[3] before the late 16th century most amber was used to make rosary beads, or turned to form vessels, as we see here. Bruges and Lübeck were two of the earliest centres for amber-turning; guilds of *Paternostermacher* were established there by 1302 and before 1360 respectively.[4] Thus although it is difficult to assign an exact date and place of origin to these cruets, they almost certainly date from before the mid 16th century, and come from Lübeck or Bruges.

Notes

[1] I am most grateful to Michael Holmes for his help on the heraldic significance of the coat of arms.

[2] Cf the two cruets in the painting of St. Eligius by Petrus Christus (New York, Metropolitan Museum), painted in Bruges and dated 1449. These are similar in form to ours, as is the engraving of a cruet in Erhard Altdorfer's *Der Glückshafen zu Rostock*, dated 1518. These are illustrated in J.M. Fritz, *Goldschmiedekunst der Gotik in Mitteleuropa*, Munich, 1982, figs 1, 41. Cf also two silver cruets for the Mass from Lübeck, dated 1518, in the Metropolitan Museum, New York, (also illustrated in Fritz, *op cit*, fig 616). A similar shaped vessel with a coconut shell as the body and an amber finial from the Abbey of Maubuisson, documented since 1463, is illustrated in *Les Fastes du Gothique: le Siècle de Charles V*, Paris 1981 (exhibition catalogue) pp 226–227, cat no 180. (The catalogue entry dates it as 14th century.) I am very grateful to Paul Williamson for drawing my attention to this.

[3] See introduction. Cf also a carved amber head of Christ set in enamel mounts in the Wallace Collection (inv 295), thought to date from the 15th century, and a head of Christ of about 1380 in the Bayerisches Nationalmuseum Munich (inv MA 2478). See *Die Parler und der Schöne Stil 1350–1400*, Cologne, 1978, II, p 710.

[4] Pelka, *Meister*, pp 7 and 8.

2   *Backgammon Board*

Königsberg, c. 1608–1647.
Amber on ebonised wood support with metal hinges.
Length 31 cm Width (open) 61.9 cm Depth 3.5 cm
A.11-1950
Given by Dr W.L. Hildburgh FSA in 1950. Department of Sculpture.

The present wooden setting was restored in 1968. Some of the clear amber panels are cracked or marked, and four missing ones have been replaced by glass. Some of the white amber reliefs are broken.

Like cat no 3, this hinged gamesboard is an example of one of the most popular secular objects for which amber was used in the late 16th and early 17th century.[1] The plain wood on its outer faces is a later restoration, and comparison with similar gamesboards suggests that originally amber was inlaid outside as well as inside, perhaps to form a chess-board and nine-men's-morris board.[2]

The open board reveals a rich display of clear amber panels over painted foil used for the points of the board, the borders of the figurative white amber reliefs under clear amber, and ovals at the centre of each of the inner upright edges, depicting profile heads of classical warriors.[3] The points and the border surrounds are painted with designs of arabesques and grotesques. On the outer upright edges are more oval clear amber panels over arabesque and grotesque designs painted on foil; these alternate with cloudy amber panels.

In the centre of the two tables are six pierced white amber reliefs set on horn, or perhaps foil painted black; these show scenes from the myth of Phaethon. In Ovid's version of the story, Phaethon was the son of Apollo the sun-god, and was presumptuous enough to ask his father to let him drive the chariot of the sun across the heavens. Apollo yielded, but Phaethon was too weak to check the horses, and plunged so near the earth that he almost set it on fire. Zeus therefore had to kill him with a flash of lightning, so that he fell into the river Eridanus. His mourning sisters wept so much that they were metamorphosed into poplars, and their tears turned to amber.[4] This last event in the story explains

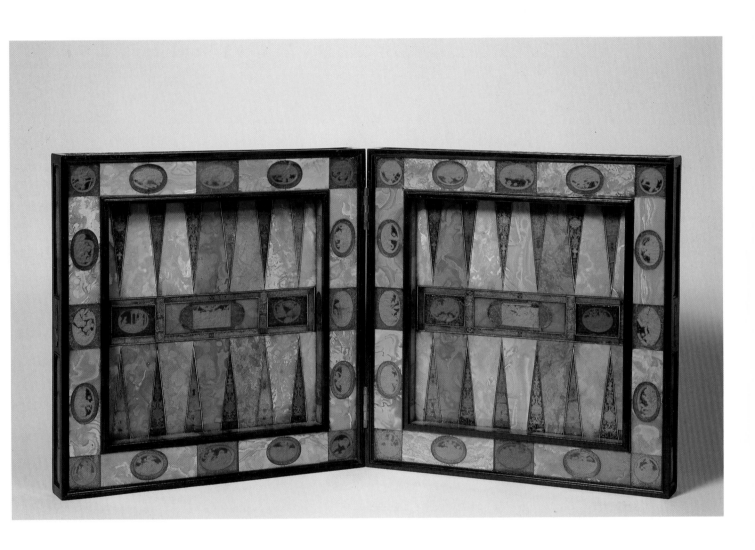

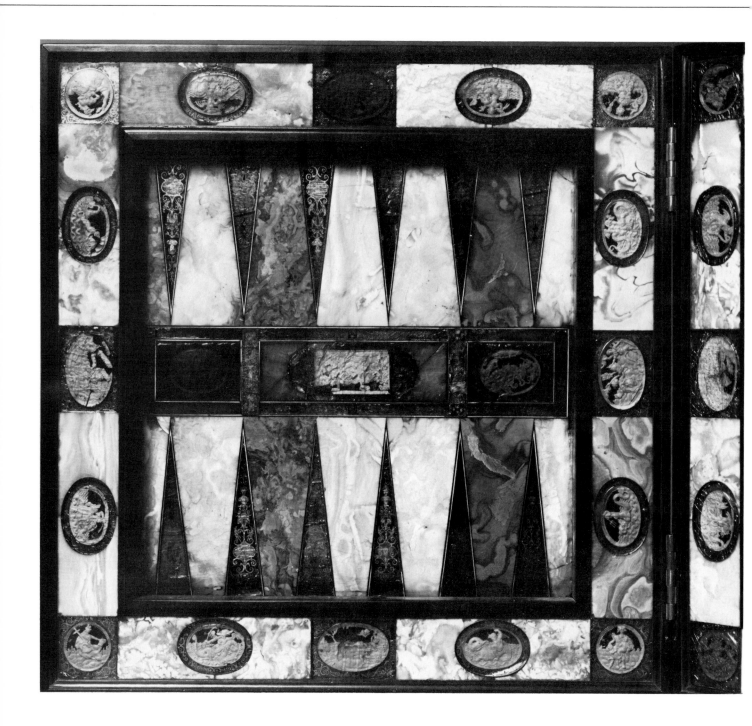

why the artist chose to depict this myth, its being a witty allusion to the origin of his material. The reliefs show (1) Phaethon approaching Apollo as a suppliant declaring his desire to drive the chariot of the sun, (2) Phaethon driving the chariot, (3) Phaethon plunging to his death and (4) Phaethon's sisters, half-transformed into trees, weeping for their dead brother. In the centre of each half of the gamesboard, interrupting the sequence of the Phaethon narrative, are two scenes also from Ovid: Midas and Silenus, and Apollo, Pan and Midas, also in white amber under clear amber.

The raised borders of the board are decorated with more pierced white amber reliefs under clear amber panels. These also depict scenes from Ovid's *Metamorphoses*: Perseus beheading Medusa, Apollo and Daphne, Pluto and Proserpine, Meleager and Atalanta, the birth of Adonis, and Apollo killing Python, as well as two indecipherable, broken scenes. Each of them is flanked by emblems of riders on fantastic sea-creatures, reminiscent of engravings by Adriaen Collaert of Antwerp (c. 1560–1618).[5] The seven Virtues are represented at seven of the corners, the eighth being occupied by an unidentified female figure.

Gamesboards such as this were produced particularly in Königsberg.[6] This one resembles a board in Kassel, thought to date from 1580–1590,[7] the decorative work on painted foil and style of the white amber reliefs being similar. Even more striking as a comparison is one now at Munich, whose reliefs not only closely resemble ours in style, but also depict the myth of Phaethon, and Apollo and Daphne.[8] As remarked above, the myth of Phaethon is eminently suitable for an amber piece, although no other examples illustrating it are known to the present writer. The Munich board has metal mounts with the mark of the goldsmith Andreas Meyer, who was in Königsberg from 1608 to 1647.[9] Although the mounts on our gamesboard have no marks, it seems reasonable to date it to within this period, and to postulate that both come from the same Königsberg workshop.

Notes

1   See Rohde, figs 49–52 and Reineking von Bock, figs 84–102. See also note 2 to entry on cat no 3. There is a similar gamesboard of about 1620 in the Kungl. Vetenskapssocieteten in Uppsala. See S. Wallin, 'Bärnstenskonst hos Kungl. Vetenskapssocieteten in Uppsala', *Kungl. Vetenskapssocietetens Arsbok*, Uppsala, 1960, pp 77 ff.
A 17th-century gamesboard with its original amber games-pieces still remaining is housed in Blair Castle. I am very grateful to Malcolm Baker for drawing my attention to it. Cf gamesboards of wood inlaid with ivory, silver and mother o' pearl made in the early 17th century, which are similar in form, eg the Hainhofer gamesboard of 1610–1617 in the Museum für Kunst und Gewerbe, Hamburg. See H. Bethe, 'Das Hainhofer Spielbrett in Hamburg und seine Verwandten', *Festschrift für Erich Meyer zum Sechzigsten Geburtstag*, Hamburg, 1957, pp 183–190.

2   See the examples cited in note 1. These missing panels could be those on the bases of cat nos 4 and 5. See entry for cat no 4.

3   Cf the engravings of Roman Emperors by the Netherlandish engraver Nicolas de Bruyn, (c. 1570–1652) (VAM E551(1–12)-1885).

4   Ovid, *Metamorphoses*, Book II.

5   See Hollstein, IV, p 203, nos 408–411.

6   Rohde, figs 51–54. Cf also cat no 3. Two gamesboards with gamespieces were made in Copenhagen by Lorenz Spengler in the mid 18th century, based on designs by Marcus Tuscher. These are now housed at Rosenborg, Copenhagen.

7   Kassel, Staatliche Kunstsammlungen, inv no BVI-217. Reineking von Bock, figs 86–93.

8   Munich, Bayerisches Nationalmuseum, inv no 79,327. Previously at Mentmore Castle. Reineking von Bock, figs 94–102.

9   Reineking von Bock, p 83.

3 Gamesboard

Amber tortoiseshell metal foil and ivory on an oak framework. Silver mounts.
Königsberg; dated 1620.
Length 39.3 cm Height (closed) 9.2 cm Width (open) 52.5 cm
Bought from G.J. Jennings in 1910; previously in the collection of the Duke of
Leeds at Gogmagog, Cambridge.
W.15-1910. Department of Furniture and Interior Design.

Some of the amber panels have been cracked and repaired; one has been replaced by
plastic. Many of the white amber reliefs are broken. Some of the tortoiseshell and ivory
panels are loose.

This richly decorated hinged gamesboard is similar to cat no 2; on the outside however, unlike cat no 2, it retains its original decoration, and is designed with a chess-board and a board for nine-men's-morris. Inside is a backgammon board.

The chess-board has squares of cloudy amber alternating with squares of clear amber placed over coloured foil. Soldiers in contemporary costume are depicted on the foil. Six small medallions of pierced white amber reliefs under clear amber probably depict scenes from Ovid, although four are broken, and hence indecipherable; the two others, which are themselves difficult to decipher, seem to illustrate *The Birth of Adonis*, and *Venus and Adonis*. The style of the reliefs and the subject matter are similar to those on cat no 2. A border of foil painted with flowers (also under clear amber) surrounds each medallion. A tortoiseshell border surrounds the perimeter of the board, with ivory strips at the edges, incised with a linear pattern and the incisions darkened with ink.

The nine-men's-morris board is decorated in a similar way to the chess-board. The lines of the game are made of strips of coloured foil decorated with foliate designs, and placed under clear amber. In the centre is a broken and indecipherable white amber relief under clear amber, possibly of a mythological scene. Six other white amber reliefs (also illegible) under clear amber are set around this central one. Each of these reliefs is bordered by foil painted with flowers also under clear amber.

The backgammon board on the interior faces is profusely ornate. Round the edges and on the inner upright edges are designs of flowers, birds, dogs hunting hare and deer, and soldiers, painted on foil and placed under clear amber panels. These alternate with panels of cloudy amber. Ivory strips and panels, incised with lines darkened with ink are set round the edges and on the inner corners. Running across the centre of the board are four Virtues: Justice, Hope, Temperance and Fortitude, painted on foil and placed under clear amber. These are not the Cardinal Virtues, and were perhaps chosen as the Virtues most to be borne in mind when gambling. In the centre of each table of the board are white amber reliefs under clear amber of indecipherable figurative scenes (the white amber having broken); each is surrounded by a border of foil painted with flowers under clear amber. The points of the board are of foil painted in red and gold, and black and gold with arabesque designs, placed under clear amber, and edged with ivory strips. Between the points are cloudy amber panels, and, in the centre of twelve of them are designs of birds in foliage on painted foil under clear amber, reminiscent of the reliefs on the amber cup, cat no 6, and probably derived from an engraved source such as the work of Adriaen Muntinck.[1] In the centre of each of the other cloudy amber panels are eight busts (perhaps portraits) of white amber under clear amber, surrounded by a flower-border of painted foil under clear amber. These are reminiscent of other early ambers of the late 16th and early 17th century.[2]

Running along the centre of the two tables is an

inscription in twelve ovals of painted foil under clear amber, reading: 'WER IN DISEM BRETT SPIL / WIL SPILEN DER MVSHABEN / DER NVMMER FIHL 1.6.2.0 / GODT GIBT FIHL MEHR AVF EINEN / TAG ALS EIN GANS KEINNIG / REICH VER MAG / DO ICH GENVCH HAT ZV GEBEN / DO KVNDT ICH BEI IDERMAN / WOL LEBEN NVICH NICHT / MEHR VERMAG SOWINSCH / MIHR NIMANDT EINEN GVTTEN / TAG GLICK VND GLAS WI BAL BRICHT DAS'. The translation of this is: 'Whoever wants to play this board game must have plenty of numbers; in 1620 God gives more in one day than a whole kingdom has. When I had plenty to give I got along with everyone; now that I no longer have, no one wishes me good day. Good luck and glass: how soon they break!'

The type of decoration on the board accords with its date of 1620.[3] At this time, Königsberg was an important centre of amber working, and the German dialect of the inscription would support this as place of origin.[4]

O. Jacoby and J.R. Crawford, *The Backgammon Book*, London, 1970, p 16. Trusted, *Album*, figs 5 and 6.

Notes

1   See entry for cat no 6, note 3.

2   Eg an amber medallion of the late 16th century of the Polish King Stefan Batorys, now in Wawel Cathedral Treasury, Cracow, illustrated in Grabowska. See also a cup and cover of the second half of the 16th century in Munich, Schatzkammer der Residenz, illustrated in Rohde, fig 16. Cf also cat nos 2, 4 and 5.

3   See entry for cat no 2, Note 1.

4   Cf Rohde, *Königsberg Pr.*, p 25. A brief discussion of the Danzig and Elbing dialects, and what does not constitute a Königsberg dialect is also given in Rohde, p 45.

4   *Two Men Fighting with a Boy and a Dog*

Base: Probably Königsberg; late 16th to early 17th century.
Figure-group: German (?); 19th century.
Base: amber and ivory on ebonised wood and boxwood core.
Figure-group: boxwood.
Height 40 cm Height of base 11.7 cm Width of base 30 cm Depth of base
18.1 cm
A.14-1950. Department of Sculpture.
Given by Dr W.L. Hildburgh FSA in 1950.

There is some surface crazing of the amber panels on the plinth, but overall the object is
in good condition.

This piece forms a pair with cat no 5. The figure-group is boxwood dowelled into the base, and only the latter employs amber, which is glued on to a wooden core.

The subject of the boxwood figure-group is obscure: two men are fighting while between them are a kneeling boy, and a dog scratching its ear. Clumsily inscribed on the top of the base is what is almost certainly a spurious inscription: 'R DE FRANCHI F.T. A.D. 1622', while on the back of an irregularly shaped block on which one figure is resting his knee is scratched 'A D 1622 MIL'. The date 1622 accords with the apparent style of the figure-group, although 'R de Franchi' seems to be an imaginary name, and the Italian provenance implied in it, and perhaps in 'MIL' for Milan, does not correspond with its apparent Northern origins: the style of the figures is similar to Netherlandish works of the early 17th century.[1] The dog scratching its ear is reminiscent of mid 16th-century South German (probably Nuremberg) bronzes, previously attributed to Peter Vischer, and ultimately based on an engraving by the Hausbuchmeister.[2] In the 19th century the Berlin Iron Foundry (active 1804–1874) produced cast iron versions.[3] The obscure subject matter, which is in part reminiscent of a *Sacrifice of Isaac* group, the incongruity of the costume, which is a combination of Mannerist-style classical armour, 17th-century buskins, and drapery of an indeterminate period worn by the kneeling boy, the overall weak carving of drapery and

anatomy, and certain telling details, such as the pose and face of the kneeling boy, suggest that this group is a 19th-century pastiche, inspired by Netherlandish sculpture of the early 17th century.[4] Germany is a likely place of origin, given the motif of the scratching dog and that the group is combined with worked amber of which there was much in Germany during the 19th century.[5]

The base however has a longer and more complex history; it displays decorative techniques typical of Königsberg in the late 16th and early 17th century. On the front and two sides, clear amber panels are set over painted metal foil and a pierced white amber relief; cloudy and opaque amber panels are set around the clear ones. Moulded ivory strips and wooden borders of inlaid boxwood and ebonised wood turned with a wave design surround the whole, while the back is ebonised wood unadorned. The painted foil panels depict (with one exception) roundels of busts of men and women, and animals in two ovals, each between a pair of roundels. They appear to be purely decorative, neither allegories nor portraits, analogous to those on a gamesboard at Kassel which is dated 1594.[6] On the front are roundels of busts of five cavaliers, a man in a feathered hat, a soldier and a woman; on the left side of the base is another bust of a woman. All are bordered with scroll designs. Like those on the base of cat no 5, these busts probably derive from an engraved source, such as the work of Virgil Solis of Nüremberg

(1514–1562).[7] In the centre of the front of the base is a small pierced white amber relief under clear amber, probably a representation of Temperance (surface crazing means that it is difficult to read clearly). On the right side of the base is another allegorical representation, here painted on foil, of a bare-breasted woman looking at herself in a mirror, with an eagle and the sun, symbolising Vanity. Round the busts and allegorical figures are borders of foil painted with arabesque designs under clear amber, and darts also of clear amber over foil (some of which is missing) decorate the sides of the base, recalling the points on backgammon boards.

There is no reason to doubt the authenticity of these panels, but in their present state they pose problems. This object, along with cat no 5, is the only known example of a boxwood figure-group being combined with an amber base, although there are a few examples of amber figures on amber socles.[8] The arrangement of the panels seems unhappy: they are crammed into narrow bands, and the roundels are set in a pattern which appears suspiciously symmetrical, yet at odds with contemporary arrangements on chessboards. Neither does the iconography of the two allegorical panels accord with the unity of themes usually found in late 16th- or early 17th-century ambers, (cf cat nos 2 and 3). All this suggests that they have been radically reorganised, and their most likely original function was quite other. They probably formed part of an early gamesboard (with panels from cat no 5), which subsequently deteriorated. Some panels from it must have been salvaged, and were given a new wooden framework in the 19th century, and the figure-group was dowelled on. The amber panels are amongst the finest early examples in the collection, close to the ones on the gamesboard at Kassel, but misappropriated at a later date. It is even possible that they once belonged on the outside of one of the gamesboards in the Museum's collection (cat no 2). These groups and the gamesboard were all given by the same donor, who perhaps in turn acquired them from one source; unfortunately the original provenance is unknown.

Notes

1. Cf the boxwood figure-group of c. 1600–1630 by the 'Master Signing XX, *The Beheading of St John the Baptist*, Victoria and Albert Museum, inv 1173-1864. See T. Muller, 'Zür Südniederländischen Kleinplastik der Spätrenaissance', *Festschrift für Erich Meyer zum Sechzigsten Geburtstag*, Hamburg, 1957, fig 5. Cf also the work of Andries De Nole, of Anvers (1598–1638), *Saint Roch*, of c. 1623, *La Sculpture au siècle de Rubens*, Musee d'Art Ancien, Brussels, 1977 (exhibition catalogue), cat no 29, p 61.

2. See E.F. Bange, *Die Deutschen Bronzestatuetten des 16. Jahrhunderts*, Berlin, 1949, cat no 84, pp 125–126. Versions exist in Bologna, Brunswick, Dresden, Nuremberg, and Paris. Cf also the bronze in the Bayerisches Nationalmuseum, Munich (inv 59/11). G. Himmelhaber (Ed), *Bayerisches Nationalmuseum, Bildführer, I: Bronzeplastik Erwerbungen von 1956–1972*, Munich, 1974, p 44. An unpublished version is in the Victoria and Albert Museum, inv A.30-1947.

3. See *Eisen Kunstguss aus der ersten Hälfte des 19. Jahrhunderts (Kleingerät und Schmuck)*, Frankfurt am Main, 1975 (exhibition catalogue), cat no 14, p 15, and W. Arenhövel, *Eisen statt Gold . . .*, Berlin, 1982 (exhibition catalogue), p 211, cat no 454. For the work of the Berlin Iron Foundry see H. Schmitz, *Berliner Eisenkunstguss*, Munich, 1917. I am very grateful to Anthony Radcliffe for his helpful suggestions on this and on the bronze parallels.

4. I am grateful to Malcolm Baker for his constructive advice on this point.

5. See Rohde, p 7.

6. Staatliche Kunstsammlungen, Kassel inv B.IV-217; Rohde, fig 49, 50. Reineking von Bock, fig 84.

7. See Franke, figs 50 ff.

8. Eg the figures of the Great Elector and his consort in the Staatliche Kunstsammlungen, Kassel (inv B.IV, 20, 21), Pelka, fig 33, Rohde, fig 144, Reineking von Bock, fig 141, the *Three Graces*, in Dresden by Christoph Maucher (Rohde, figs 147, 148) and the two figures of *Jael* and *Judith* also by Christoph Maucher in the Museo e Galleria Estense Modena, (inv 839, 840) see Trusted, *Pantheon*.

5   *Three Men Fighting*

Base: probably Königsberg; late 16th to early 17th century.
Figure-group: German (?); 19th century.
Base: amber and ivory on ebonised wood and boxwood core.
Figure-group: boxwood.
Height 40 cm Height of base 11.7 cm Width of base 30 cm Depth of base 18 cm
A.15-1950
Given by Dr W.L. Hildburgh FSA in 1950. Department of Sculpture.

The lower right arm of one man and the end of the finger of his left arm are missing; the tips of the fingers of the kneeling man are missing, and his left arm is cracked across. The sword of the man on the right is missing, and the drapery of this figure is also chipped. Some surface crazing on the amber panels on the base is evident, and some are slightly cracked.

Although there are a few minor differences, this piece forms a pair with cat no 4. The subject here is similarly obscure: two men in feathered caps, but wearing apparently classical costume, are in combat, while a third man, more simply dressed, kneels between them, apparently suppliant to and protected by the one on his right. The poses are stiff, and the carving weak.

The base is bordered by ebonised wood moulding and inlaid boxwood, and moulded ivory strips. On the front of the base, painted on foil under clear amber panels, are busts of two soldiers, two Eastern potentates, four cavaliers, and two small ovals of a horse and a lion. The busts are bordered by scroll patterns. On the right side of the base is the bust of a cavalier, and on the left, a soldier, both similarly painted on foil under clear amber with decorative borders of foil painted with scroll and crescent patterns, probably derived from a source such as Virgil Solis of Nüremberg.[1] In the centre of the front is a white amber figure, perhaps representing Charity, but this is difficult to see because of surface crazing. The darts on cat no 4 are not present here, and instead, on the sides, are mosaics of clear and opaque amber panels. The back of the base is ebonised wood unadorned. The amber panels on the base probably come from the same gamesboard as those on cat no 4, perhaps the one now in the Museum's collection (cat no 2). The figure-group however is probably a 19th-century pastiche of 17th-century Netherlandish sculpture, and must be by the same artist (or faker) who carved the group on cat no 4.[2]

Notes
1 See note 7 in entry for cat no 4.
2 See entry for cat no 4 for detailed references.

6   *Two Handled Cup*

Königsberg; mid 17th century.
Amber, gilt brass mounts.
Height 11.7 cm Diameter of base 7 cm Diameter of cup 11 cm
659-1904
Bought from J. Jackson Esq in 1904. Department of Sculpture.

Three of the seven amber panels of the cup have been broken and repaired. Slight chips and marks occur on the other panels. This damage and the ill-fitting of some panels indicate that all of them were at some time removed and fitted back into their present position with a new metal lip-band, which looks like a recent restoration, as do the metal stem and circular metal base beneath the foot. The handles have been re-soldered on to the band, and one of them is positioned in such a way as to block one of the decorative panels. When bought, a gilt brass rococo border was attached by screws to the foot, to which it had been added. This was removed.

This cup would have been a prime candidate for a *Kunstkammer* collection of precious objects. The elegant arabesque gilt handles are both set with an oval amber boss; their form parallels those of mid 17th-century silver vessels.[1] The figurative carvings depict trophies of animals and fruit. The animals represented are: a pelican, a stag, an eagle, a squirrel, a fox, a swan, and a rabbit. The object was intended to be picked up and studied from all angles, for the use of clear amber means that the carvings can be viewed from both inside and out. The foot of the cup is adorned with three metal panels engraved with acanthus leaves and three clear amber panels carved with fantastic sea-creatures. A number of similar cups can be compared in general form with this one, all probably from Königsberg and dating from about 1600 to 1650.[2] The parallel examples have amber, not metal stems, suggesting that originally 659-1904 had an amber stem too. An even closer comparison can be made with an amber cup in the Cluny Museum,[3] thought by Rohde to be from Königsberg, and to date from the first quarter of the 17th century. It is not only the same shape, but the carvings of its panels correspond so closely to ours that one is tempted to assign both to the same workshop. The exact engraved sources for the carvings have not been traced, but they resemble the engravings of the Netherlandish engraver Adriaen Muntinck (active in Groningen and Amsterdam c. 1597-c. 1617); his designs for goldsmiths' ornaments were published by C.J. Visscher after 1597.[4] Despite Rohde's dating, the handles of this piece imply that it is of the mid 17th century rather than earlier.

Königsberg was the sphere of activity of the important early 17th-century amber worker, Georg Schreiber (active c. 1617–1643). The carvings of fruit and scrollwork on tankards and caskets attributed to him parallel the rather broader carving on this cup,[5] and it is likely that Königsberg workshops influenced by his work produced many such cups to grace the aristocratic *Kunstkammer* collections.

Rohde, p 36, fig 81.

Notes

1   Cf handles on Scandinavian silver vessels, eg cat nos 33, 92 in R.W. Lightbown, *Catalogue of Scandinavian and Baltic Silver Victoria and Albert Museum*, London, 1975.

2   These are a cup in Vienna, Kunsthistorisches Museum, inv no 3550, Rohde, fig 84, Reineking von Bock, fig 152; a cup formerly in the Berlin Schloss Museum, Rohde, fig 85; a cup and cover sold at Sotheby's, 13 April 1978, lot 64, the same piece exhibited at Colnaghi's in 1981; see A. Gonzales-Palacios and L. D'Urso, *Objects for a 'Wunderkammer'*, London, 1981, cat no 129, pp 272–273. The carving corresponds with that on a beaker in Vienna, Kunsthistorisches Museum, inv no 3543, Rohde, fig 91, Reineking von Bock, fig 151, and a chalice in the Bayerisches Nationalmuseum, Munich (inv R2758).

3　Cluny Museum, Paris, inv no Cl 659. Rohde, figs 82, 83.

4　Victoria and Albert Museum, inv E892ff-1963. See U. Thieme and F. Becker, *Allgemeines Lexikon der Bildenden Künstler von der Antike bis zur Gegenwart*, Leipzig, 1907–50, XXV, p 278. Possibly the same source was used for the ivory carvings on the inside of a late 17th-century casket in Kassel. See Pelka, fig 57.

5　For examples of work attributed to Schreiber, see Pelka, figs 50, 66, Rohde, figs 68–71 and Reineking von Bock, figs 113–117, 119–121. See also O. Pelka, 'Zum Werk des Bernsteinmeisters Georg Schreiber in Königsberg', *Pantheon*, XVII, 1936, pp 27–29.

7 Two Handled Bowl

Königsberg, mid 17th century.
Amber; silver gilt mount and later nickel-coated brass mount.
Height 7.5 cm Diameter of rim 14.5 cm
A.9-1950
Given by Dr W.L. Hildburgh FSA in 1950. Department of Sculpture.

Three of the larger figurative panels and three smaller panels are badly cracked, and have been repaired. Surface crazing and minor marks occur on the other panels. The amber handles have been removed and later glued back in the wrong positions; one of them (made-up, of recent origin) has obliterated one of the figurative panels, and the middle band of metal (nickel-coated brass) is a later replacement. All these features suggest that the bowl was dismantled and reassembled at some time. The three ball-feet are scratched and chipped.

Like the amber cup (cat no 6), this bowl was almost certainly intended for a *Kunstkammer* collection. The iconography of the figurative carvings is purely secular; the twelve main panels of clear amber depict the labours of the months, represented by agricultural labourers in contemporary 17th-century costume performing seasonal tasks, except for January, who is represented by a gentleman with a lute, and August, who appears as a vine-wreathed Bacchic figure; these are probably based on German or Flemish engravings, such as those by Virgil Solis or Jost Amman.[1] Inside the base is a circular pierced relief of white amber on a black painted foil backing, with a circular border of gold painted foil, under clear amber, depicting quarter-length profiles of a man and woman gazing at each other, the man holding a goblet.[2] The clear amber handle (along with the restored one) is broadly carved in an arabesque form, and the bowl rests on three ball-feet, with gadrooning on the lower smaller clear amber panels to which the feet are fixed. Both the handles and the ball-feet echo the forms of contemporary Scandinavian and Baltic silver.[3]

The shape of the bowl has parallels with works in Vienna, Budapest, Florence and Rome,[4] although all of these are more deeply carved and display more elaborate compositions than ours. They are attributed by Rohde to the circle of Jacob Heise of Königsberg, (active 1654–1663).[5] The figures of the months on our bowl are closer in style to the low relief figures on tankards made in the circle of Georg Schreiber, also of Königsberg, (active c. 1617–1643).[6] This bowl probably dates from the period of Jacob Heise's activity: its form is analagous to the bowls attributed to his circle, while the carving continues the tradition begun by Schreiber.

Rohde, p 38, fig 110.

Notes

1 See Franke, plate 35, e 2–5. Cf also the engravings from Jost Amman's *Kunstbüchlein . . . vieler geistlicher und weltlicher hohes und niederstands Personen* of 1599, and Hans Weigel's *Trachtenbuch* of 1577, on which are based the decorations of a silver double goblet made in Zurich c. 1600, in the Schweizerisches Landesmuseum, Zurich. See A. Gruber, *Weltliches Silber Katalog der Sammlung des Schweizerischen Landesmuseum Zürich*, Berne, 1977, p 58, cat no 60. I am grateful to Dr Hanspeter Lanz for enabling me to see this object.

2 A similar relief (the bust of a woman carousing) of white amber under clear amber is on the inside of the lid of a tankard in the British Museum, dated 1659, Rohde, fig 119 (the relief is not illustrated). A polygonal amber box in Malbork Castle Museum also contains such a relief on the inside.

3 See R.W. Lighbown, *Catalogue of Scandinavian and Baltic Silver Victoria and Albert Museum*, London, 1975, cat nos 33 and 130. Cf also the handles of the two-handled cup attributed to Georg Schreiber in the Malbork Castle Museum, illustrated in Grabowska.

4 Rohde, figs 139–142

5 Rohde, p 42

6 Rohde, figs 108, 109, 114, 122, etc.

8   Two-Tiered Casket

North-East Germany, perhaps Königsberg. Mid 17th century.
Amber, ivory, mica or horn backing, silver-coated brass mounts; later
restorations of ribbon, and glass.
Height 25.3 cm Length 20 cm Depth 13.1 cm
269-1875
Given by Miss E.M. Dorrell in 1875. Department of Sculpture.

There is some surface cracking on the amber, and small pieces of amber are missing from
the lower tier. The central pierced ivory relief inside the base may have originally been
under a clear amber panel: a later glass panel replacement now covers it.

The casket is surmounted by ivory figures of two dancing putti on half-moons. At each corner of the side of each tier is a bare-breasted female herm of white amber with auricular masks forming the lower part of each; these recall the caryatids on a 16th-century doorway in Königsberg Castle (now destroyed).[1] Inside the base are three ivory reliefs: an octagonal pierced ivory relief, laid on mica or horn, of a shepherd reclining in a landscape, flanked by two oval (unpierced) reliefs of a shepherdess and a shepherd. (The former, pierced relief would probably originally have been seen through clear amber: compare cat nos 2 and 7 and see above.) The casket rests on four amber segmented double bun feet waisted with a band of amber. Four clear amber balls on the lower tier and four cloudy ones on the upper tier are fixed by a metal tang covered in gold foil drilled through the centre with finials of ivory rosettes.

This casket has no wooden support structure: the amber slabs have been dowelled and glued together, and the decorative female herms perform an essential function in concealing the joins at the corners. On the inside, ribbons fixed to metal mounts help support the open lid of the lower tier; hinges and locks are fitted on both tiers. At the front, sides and back of both tiers and in the base of the lower tier, amber slabs have been faceted on one side and semi-spherical and oval cavities gouged out of the other side, to create decorative effects of light, particularly when the casket is open.

Williamson, p 173.

Rohde, figs 211 and 212.

W.L. Hildburgh, 'An Amber and Ivory Altar', *Apollo*, XXX, 1939, pp 208–213.

K. Aschengreen-Piacenti, H. Honour and others, *Ambre Avori, Lacche, Cere, Medaglie e Monete*, Milan 1982, p 28, fig 2.

A. Burton and S. Haskins, *European Art in the Victoria and Albert Museum*, London, 1983, p 76.

Notes

1 The wall-monuments at the Oliva Monastery near Gdansk are particularly close parallels.

2 See Hollstein, XVII, p 250.

3 *The Infant Christ and St. John the Baptist*, Landesmuseum, Kassel, Rohde, fig 249. Dated by Rohde to the early 18th century, when many of the ambers at Kassel were being carved. Cf also cat no 22.

4 Cf the amber Madonna now at the Shrine of the Black Madonna, Jasna-Gora, Czestochowa, Poland. Illustrated in Grabowska, given to the Shrine in 1604, and probably dating from the late 16th century. See also Z. Rozanow and E. Smulikowska. *The Cultural Heritage of Jasna Gora*, Warsaw, 1979, p 97.

5 See Pelka, fig 55, Rohde, fig 23. This decoration can be seen on numerous 17th-century caskets. See Pelka, figs 56, 58, 59, Rohde, figs 159, 160, 163.

6 Malbork Castle Museum, illustrated in Grabowska, and attributed by her to the workshop of Christoph Maucher, although the present writer cannot reasonably attribute it to him. As stated in the entry for catalogue no 6, the source has not been identified, but the work of Adriaen Muntinck provides close analogies. Such swags (ultimately derived from ancient Roman sculpture) however can be seen in the work of Italian artists from the 15th century onwards, such as Mantegna's altarpieces, and the works of Agostino Carracci (see Bartsch, XXXIX, p 293, 273–274).

7 See F.W. Hollstein, VI, p 250.

8 Examples are housed in the Victoria and Albert Museum (inv E891ff-1963). See also entry for cat no 6, note 3.

9 See Franke, f102–106, g49–75.

10 See W. Hütt, *Albrecht Dürer Das Gesamte Graphische Werk*, Munich, 1970, II, p 1528.

11 Rohde, p 55 and fig 210.

12 See K. Aschengreen Piacenti, 'Due Altari in Ambra al Museo degli Argenti', *Bollettino d'Arte*, III–IV, July–December, 1966 pp 163–166.

13 Illustrated in Grabowska. Previously sold at Christie's, London 23 May 1966, Lot 82.

10 Shrine

Amber on wooden core. Ivory reliefs on horn backing. Metal hinges and locks.
North East German, probably Danzig. Mid 17th century.
Height 25.8 cm Length 17 cm Depth 10.4 cm
270-1875. Department of Sculpture.
Given by Miss E.M. Dorrell in 1875.

During conservation work on the shrine in 1982, a small ivory group of the Crucifixion surmounting it was removed, as it was undoubtedly a later addition, replacing the lost original figure of the Virgin, of which survivies only the half-moon (with the face of the man-in-the-moon) once at her feet.[1] One of the ivory swags on the second tier is missing, and there are slight cracks and surface crazing on some of the amber panels; otherwise the object is in good condition.

This highly worked precious object was probably dedicated to the Virgin; its narrative scenes depict the events around the birth of Christ, and the missing crowning figure of the Virgin would have been the most prominent feature.

The construction of clear and cloudy amber panels and ivory reliefs glued to a wooden core is the same incrustation technique seen in cat nos 9 and 11. The three tiers rest on four ball feet of clear amber; panels also of clear amber, faceted above, and with semi-spherical cavities gouged out on the underside, placed over foil, decorate the lowest tier and back, and a clear amber panel carved in a star-shape and laid over foil is set over the central door. Other clear amber panels have been carved in intaglio with floral designs on the underside and have again been placed over foil; a small intaglio scene of *The Flight into Egypt* over foil is set beneath the door. This little scene complements the other intaglio carving, on the inner face of the door, of *The Journey of the Magi*, the style of both being simple, even crude. In a variation of the églomisé technique, the central clear amber panel at the back of the shrine has been painted on the underside in gold with the symbol for Christ, 'IHS' and a cross, with flowers at each corner; this has then been laid over foil. The clear amber arabesque-shaped pieces dowelled and glued on to the edges of each tier, enhancing the outline of the shrine, and the eight clear amber balls on the second and third tiers, recall the crucifix (cat no 11), and

suggest a possible common origin. The overall form (like cat no 9) is reminiscent of tomb monuments surviving in and around Gdansk, particularly those at the Oliva Monastery, indicating that the shrine probably originates from Danzig.

One (once two) ivory swags and pierced ivory reliefs on horn adorn the front of the shrine, the latter depicting, on the first tier, *The Rest on the Flight into Egypt* and *The Baptism*, on the second tier, *The Annunciation*, and on the upper tier a vase of flowers. Inside the shrine is a slightly larger ivory relief also on horn set into the back of the 'IHS' panel mentioned above of *The Adoration of the Shepherds*. The style of all these reliefs seems closer to those on the gamesboard (cat no 2) than to the more meticulous ones on the crucifix (cat no 11) suggesting that the shrine is earlier than the latter. The inner sanctum of the shrine, perhaps intended for a relic (a statuette would have blocked the view of the ivory *Adoration of the Shepherds*), is richly decorated, and the fall of light through the open door, and from the clear pale orange amber 'windows' (scratched with a pattern to imitate diamond window glass-panes) at the back give this small area a reflected golden glow. The floor is made of transparent horn, with gold coloured foil applied to the underside in alternate lozenge shapes to imitate receding tiles.

Like the crucifix (cat no 11), Rohde believes that this was an export article from Königsberg, Danzig or

Stolp, one of many made in the late 17th and early 18th century.[2] As suggested above, the form indicates the shrine is from Danzig, and the iconography implies that it was dedicated to the Virgin, which would be more in accordance with Catholic Danzig than Protestant Prussia. Although it is difficult to give a precise date, the ivory reliefs suggest an earlier rather than a later period.

Rohde, p 55, fig 224.
Trusted, *Album*, fig 7.

Notes

1   See Rohde, fig 219, also the Madonna and Child at Jasna Gora, illustrated in Grabowska, a figure of the Virgin sold at Sotheby's 23 January 1961, lot 162, and one in Hamburg Museum für Kunst und Gewerbe inv no 1923, 88, for examples of the type of this lost figure. I am very grateful to the restorer, Carol Galvin, for her information about the construction of the shrine.

2   Rohde, p 55.

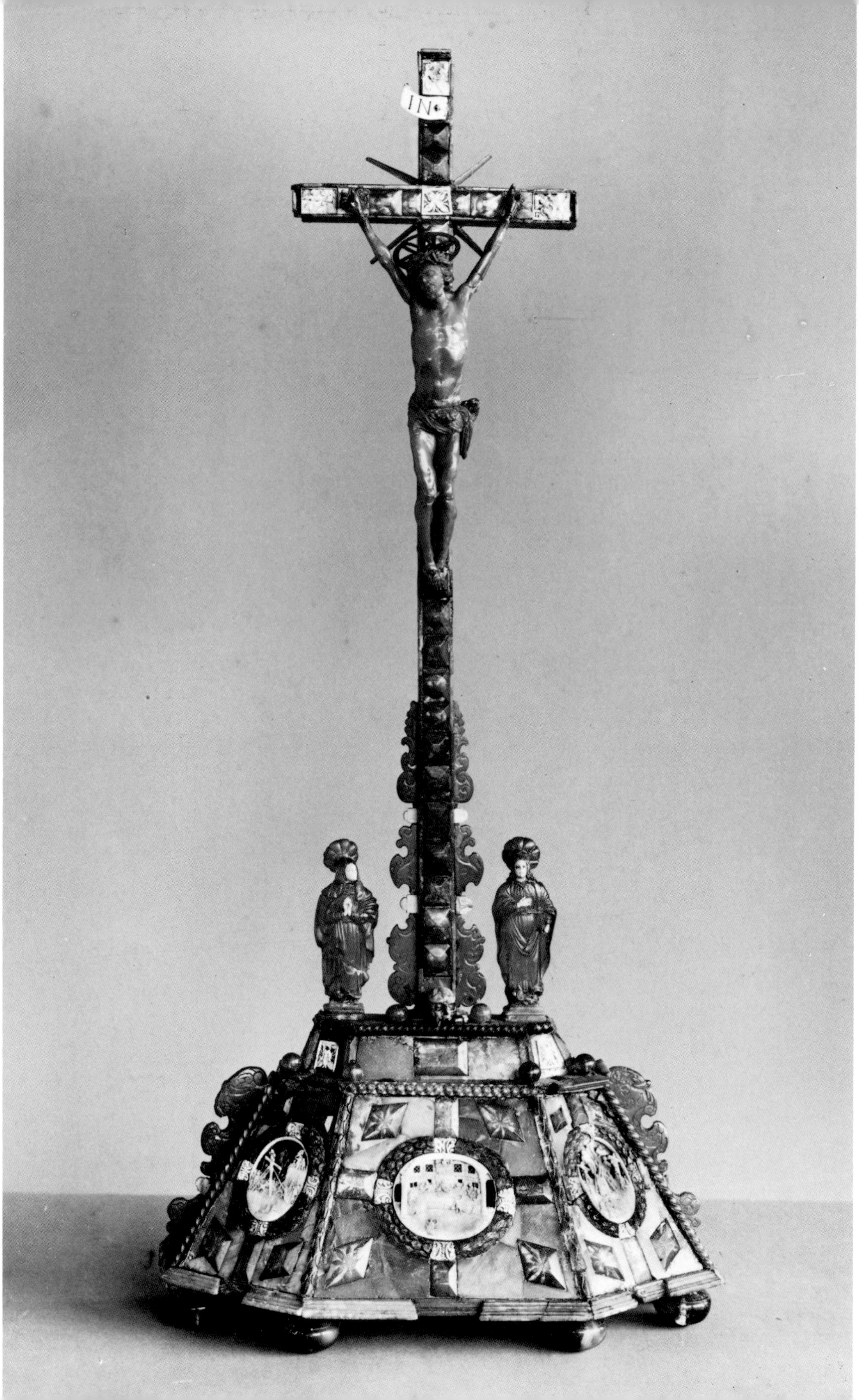
IN

11   Crucifix on Socle

North East Germany, probably Danzig. Late 17th century.
Amber, ivory reliefs set on mica backings, on wooden core. Silver gilt halo on
Christ. Later plastic restorations.
Height 66.7 cm   Width of socle 30.6 cm   Depth of socle 12.4 cm
4064-1856. Department of Sculpture.
Bought. Provenance unknown.

The figure of the Virgin is broken and repaired. The figure of Christ is broken at the
waist, knees and shoulders; the left hand is missing except for the thumb. St. John the
Evangelist's halo is broken off. One of the points of the star of clear amber on the cross at
Christ's head is broken and repaired. Some amber panels are missing; some have been
replaced by plastic, such as one of the six amber balls on the socle. Two of the small
rosettes on the balls are missing. Painted wooden pieces replace missing amber on the
sides of the cross. A photograph of 1927 (see illustration) shows remains of an ivory scroll
above Christ's head inscribed 'I.N.' (the 'R.I.' half of the scroll was already missing); this
scroll is now completely lost, although remains of glue and a screw survive. The cross
was once decorated with pierced ivory reliefs of the four Evangelists; only three remain;
St. John the Evangelist (top), St. Luke (at Christ's right hand), and St. Matthew (at
Christ's left hand). The missing St. Mark has been replaced by a relief of fruit or
vegetable form (perhaps a gourd?). The reliefs have been rearranged, concealing this loss
rather clumsily: the vertically composed relief of St. Matthew, which has been placed on
its side to fit into its present position, was probably below Christ's feet, and was replaced
by a panel of faceted amber. The missing St. Mark must have once occupied the space at
Christ's left hand and would have been horizontally composed.[1]

Christ is suspended from the cross, metal nails fixing
His hands and feet. At the foot of the cross, on a smaller
scale than the figure of Christ, are the figures of the
Virgin and St. John the Evangelist. A skull and bone
(originally two bones) lie between them. On the front
of the upper tier of the socle are two pierced ivory
reliefs of fruit or vegetable forms (see above for the
placing of other ivory reliefs). On the front of the
lower tier are three oval pierced ivory reliefs: the *Last
Supper*, flanked by the *Elevation of the Cross* and the
Deposition. At the back of the socle is an octagonal
pierced ivory relief of the *Resurrection*. The socle rests
on six clear amber bun feet.
   The altar is constructed from rectangular and
rhomboid amber slabs glued on to a wooden core, a
technique known as incrustation, introduced in the
mid 17th century.[2] Carved pieces of clear amber have
been inserted to protrude from the sides of the cross

and socle, fixed by glue and dowelling. The sides of the cross have been carved with tendril motifs; other surface carving of stylised leaves is seen on the pipe-shaped amber strips dividing the planes of the socle, and in the laurel wreath borders around the three oval ivory reliefs; four white amber motifs adorn each border. On the socle, some faceted amber panels are carved on the underside with stylised floral motifs, and placed over gold-coloured metal foil; damage at the back of the socle has laid bare the foil. On the cross, semi-spherical cavities have been gouged out of the back of some of the amber panels which are placed over gold-coloured metal foil. Below the skull, and at the back of the socle the foil is painted black, with some of the black paint scratched away to create a gold design, then a clear amber panel has been super-imposed; a technique which was used from the 16th century onwards.[3]

Two small pierced ivory reliefs of flowers under clear amber are on the reverse of the cross. Four clear amber balls on the lower tier, and two on the upper tier of the socle have been drilled through with metal screws (the top two are missing), and glued on to the structure. The top of each screw was mounted with a white amber rosette (only one survives). The faces (with painted black dots in the eyes) and hands of the Virgin and St. John the Evangelist, and one foot of the latter, are of white amber set into the clear amber of their bodies. This technique dates back to amber figures of the 15th century onwards.[4]

The present crucifix suggests parallels with similar crucifixes and house-altars;[5] Rohde believes this type of work was made in Königsberg, Danzig and Elbing for export,[6] and although there are many analogous pieces, 4064-1856 is of considerably higher quality.[7] The shape of the socle recalls the candlestick bases in the Kunstgewerbemuseum Berlin, thought to be from Danzig, and to date from the end of the 17th century,[8] while the balls on the socle parallel those on cat no 10. The drapery of the figures recalls the figures from an altar in Vienna dated about 1645.[9] The ivory reliefs on the front of the socle are of meticulously fine quality; their compositions may derive from Flemish or German 16th-century engravings, such as those by Master S, called Sanders Alexander van Brugsal of Antwerp (active 1505 onwards; d. 1554), and Albrecht Altdorfer of Regensburg (b. 1480; d. 1538).[10] The *Resurrection* on the back is in a broader style, but must

The Elevation of the Cross

The Last Supper

The Deposition

The Resurrection

be by the same hand. The richly decorated back suggests that such a piece would have been closely examined from all angles.[11] The high quality of the carving, the use of silver gilt for the halo, and the various skilled applications of amber indicate that this piece originated in a thriving centre for amber workers; Danzig is likely, given the similarity with the Berlin candlesticks and with cat no 10. Probably the artist, as was common, was skilled in both amber and ivory carving, although the possibility cannot be excluded that an amber-worker collaborated with an ivory-worker.

Rohde, fig 223 and p 55.

Notes

1 The original positions of the Four Evangelists probably corresponded with those on other amber crucifixes. Cf an altar in the Altona Museum Hamburg (inv no 1976/205) illustrated in H. Lungagnini, 'Ostdeutsche Kunst und Kultur', *Altonaer Museum in Hamburg*, Schausammlungen, X, Hamburg, 1976, and Rohde, figs 218 and 222. An amber crucifix in the Nationalmuseum, Stockholm (unpublished), provides another parallel.

2 Rohde p 54 and see introduction.

3 See introduction.

4 Cf Rohde, figs 1 and 2, the Madonna at Jasna Gora, in Poland, dated 1604, illustrated in Grabowska (also Z. Rozanow and E. Smulikowska, *The Cultural Heritage of Jasna Gora*, Warsaw, 1979, p 97), and a figure of the Virgin sold at Sotheby's, 23 January 1961, lot 162.

5 Rohde figs 213–218 and Reineking von Bock, figs 170–174.

6 Rohde, p 55.

7 Cf the works listed in note 5, and unpublished works at Lübeck, St. Annen-Museum, inv nos 282 and 1912-137. Unpublished crucifixes are also in the collection of the Duke of Northumberland at Syon House, Middlesex, and in the Nationalmuseum, Stockholm.

8 Inv no 1981/4a,b. Reineking von Bock, fig 153. Previously in the Mentmore Collection, sold at Sotheby's in May 1977.

9 Schatzkammer der Kapuziner inv no 274. Reineking von Bock, figs 175–177.

10 Hollstein, XIII, pp 123–136. Cf the woodcut by Altdorfer of the *Elevation of the Cross*, in Bartsch, p 135.

11 Cf cat no 9.

12 *The Judgement of Paris* see frontispiece

Christoph Maucher (b. 1642; d. after 1701).
Danzig. About 1690–1700.
Amber on wooden core. Modern plastic restorations on socle.
Height 19.7 cm Width 19.5 cm Depth 11 cm
1059-1873. Department of Sculpture.
Bought from Mrs Matthew Marshall in 1873.

This figure-group probably originally formed the decorative crowning piece of an amber cabinet (see below). The figures of Paris and Venus are seated with Cupid between them, in front of Juno, Mercury and Minerva. The right arm of Juno, the tips of the fingers of her left hand, and the top of her crown are missing. Minerva's and Venus's right index fingers are missing, and the top of Minerva's lance is lost. The putti medallions on the socle are slightly chipped. Some amber panels on the socle are missing, some replaced by plastic. Surface cracking has occurred on a number of the clear amber panels on the socle.

This is one of the finest ambers in the collection: the classical story is portrayed with perhaps unintentional peasant humour and sensitive, yet chunky, carving. It was first attributed to Christoph Maucher in 1933,[1] and since then this attribution has been generally accepted. Maucher (b. 1642; d. after 1701) was an amber and ivory worker active in Danzig from about 1670 onwards who undertook commissions from the Brandenburg Court in Berlin.[2]

The freestanding figure group is carved from one large piece of opaque red amber, with three additional pieces of similar amber carved with flowers and grass cut to fit round, and glued on to the base of the group. The socle on to which the group has been glued and dowelled, is formed of amber slabs glued to a wooden core. In the centre of each concave side of the socle is a medallion of a putto in acanthus leaves: three of these are of white amber, the fourth, at the back, is of the same type of red amber as the figure group above. On either side of the putto medallions at the front and back is an oval landscape scene; these four are each formed by a slab of clear amber, carved on the underside with the depiction of the scene, placed over a sheet of gold coloured foil. Each slab is held in place by two ivory headed nails. These imaginary landscapes suggest an Oriental influence, seen for example in the trees, boat and architecture of the scene below Paris. This combining of gold-coloured foil with clear carved amber gives an exotic, shimmering effect not unlike Oriental lacquer work. Chinoiserie scenes can be seen on several surviving ambers, notably cat no 18, as well as a cabinet at Brunswick,[3] and casket in Uppsala[4] and panels (now lost) from the amber throne made for Emperor Leopold I.[5] The Oriental influence may have been inspired by engravings.[6] Classical influences too can be seen in the architecture of the scene below Venus; such a combination of influences illustrates Maucher's unclassical juxtaposition of different sources. Elsewhere on the socle are floral designs formed likewise by panels of carved clear amber placed over foil. Some of the amber panels have had semi-spherical holes gouged out of the underside and have then been placed over gold foil.

The slightly awkward but charming tone of the piece is typical of Maucher, and similarities can be seen in other amber works attributed to him: details such as the three deities behind intently watching and gesturing, the dog raising its head to look at Venus, the owl at the corner beating its wings, the highly unclassical proportions of the figures, and the downward turned eyes and half-opened mouths of the faces are all hallmarks of Maucher's style.[7] The two seated figures of Paris and Venus, their legs set wide apart and feet planted firmly on the ground recall in

particular the figure of the Emperor Leopold I, in Maucher's one signed work, the ivory monument to the Emperor and his son, King Joseph I, in Vienna, dated 1700.[8] The surviving works attributed to him to not readily suggest a stylistic chronology, but the high quality of this piece implies that Maucher had fully mastered his art, and the close similarity in pose between figures on this and the Leopold monument suggests a date towards the end of the 17th century.[9]

The group and its socle seem at first to be a complete work of art, but comparison with an amber cabinet sold at Sotheby's in 1961 and now in Malbork Castle Museum strongly suggests that our group originally formed part of a similar one.[10] Surmounting the Malbork cabinet is a figure group on a sloping socle, analagous to the present piece. The cabinet probably also comes from Danzig, and dates from the late 17th century, although the figure group is smaller than ours and not in Maucher's style. In the base of 1059-1873 are six drilled holes: one at each corner and two near the middle. The socle was probably therefore once fixed to a larger object, almost certainly an amber cabinet.[11]

E.F. Bange, *Die Bildwerke in Holz, Stein und Ton, Kleinplastik*, Berlin and Leipzig, 1930, p 105.
I. Baker, 'Old Amber', *The Connoisseur*, LXXXVIII, December 1931.
Ibid, 'The Story of Amber', *The Antique Collector* 1 February 1951.
R. Verres 'Der Elfenbein- und Bersteinschnitzer Christoph Maucher', *Pantheon* XII, 1933, pp 244 ff.
Rohde, p 44 and fig 146.
J. Rasmussen and others, *Barockplastik in Norddeutschland*, exhibition catalogue, Musuem für Kunst und Gewerbe, Hamburg, 16 September– 6 November 1977, cat no 141, pp 430–432.
Trusted, *Pantheon*, fig 8.
Trusted, *Album*, fig 10.

Notes

1 Verres, p 244.

2 See Verres, and C. Theuerkauff, 'Kaiser Leopold, in Triumph wider die Türken . . . Ein Denkmal in Elfenbein von Christoph Maucher, Danzig', *Hamburger mittel- und ostdeutsche Forschungen*, IV, 1963, pp 60–93, and Trusted, *Pantheon*.

3 See Rohde, figs 292, 294, The landscapes showing Chinoiserie influence are not however visible in these illustrations.

4 Wallin, pp 106–109.

5 Baer, p 123.

6 Cf Baer, pp 122–124.

7 Cf the two figure groups at Berlin, Staatliche Museen Preussischer Kulturbesitz, Skulpturengalerie, *Perseus and Phineas*, inv no 858 and *The Judgement of Paris*, inv no 859, (Rohde figs 149–152), the *Three Graces* in the Grünes Gewölbe, Dresden, inv no III, 64 (Rohde, figs 147–148). This latter work is attributed by Rohde, along with the present piece and cat no 13, to the Master of the Königsberg Judgement of Paris. See also G. Weinholz, *Grünes Gewölbe, Gefässe und Gerate aus Bernstein*, Dresden, p 41, (cat no 18). *Judith* and *Jael*, in Modena, Galleria Estense, inv nos 839P and 840P and *Dido and Cleopatra* in the Kunsthistorisches Museum, Vienna, have been recently attributed to Maucher, (see Trusted, *Pantheon*). Two lost works probably also by him are *Cimon and Pero*, previously in the collection of Ole Olsen, Copenhagen, (Herman Schmitz, *Die Kunstsammlungen Ole Olsen I*, Copenhagen 1925, cat no 984, plate 89), and a relief of *Venus and Adonis* previously in the collection of Georg Hirth, Munich (Collection of Georg Hirth, Munich 13 June–21 June 1898, II, p 71, cat no 1216). See also the entry on the other amber by Maucher in the Victoria and Albert Museum, cat no 13.

8 Vienna, Kunsthistorisches Museum. See Theuerkauff, *op cit*, in note 2. See also *Barockplastik in Norddeutschland op cit*, p 432.

9 The Berlin *Judgement of Paris* has been in the Elector's *Kunstkammer* since 6 December 1690. See Bange, *op cit*, p 105. It is probable that Maucher died in the early years of the 18th century, rather than after 1721 as proposed in Trusted, *Pantheon*. Elzbieta Mierzwinska has pointed out in correspondence that the Latin passage in that article does not necessarily imply Maucher was alive in 1721.

10 Sotheby Parke Bernet, London sale 23 January 1961, lot 174. Illustrated in Grabowska, and there attributed to the workshop of Christoph Maucher.

11 *The Judgement of Paris* in Berlin (see note 2), also shows signs on its base of having been once fixed to something, but it lacks a socle. Perhaps it too once surmounted an amber casket. Dr Christian Theuerkauff has suggested this verbally.

13   The Judgement of Paris

by CHRISTOPH MAUCHER (b. 1642; d. after 1701).
Danzig; late 17th century.
Amber
Height 12.8 cm Width 12 cm
A.61-1925. Department of Sculpture.
Bought at the Humphrey W. Cook Sale (a part of the collection formed by
Sir Francis Cook, Bart) (Christie's, London, 10 July 1925, Lot 460).

The amber is slightly chipped; Juno's crown is broken; the background trees probably
had more foliage than at present; there is an abrupt break. Otherwise the piece is in good
condition. Six holes drilled at irregular intervals on the back indicate that at one time the
relief was fixed to a backing (cf cat no 22). Scratched on the back is the inscription
'Carrolus Maruti Ex! 1621'. This is certainly a later addition. The present writer can find
no reference to this artist; the style of the piece could not be of the early 17th century,
and immediately suggests Maucher. It is interesting but probably coincidental that the
initials 'C M' of Carrolus Maruti are also those of Christoph Maucher.

Paris, half-nude, is seated on a grassy mound under a
tree with his arm round Venus, also wearing only
slight drapery and a necklace; she half-stands, half-
leans against his left leg, about to receive the prize of
the apple. On their left are Juno with her peacock and
Minerva with her owl, and seated at the front, with a
dog, is Cupid. The relief is carved from one piece of
amber, with the additions of the owl, which is keyed
into the back, and some of Venus's drapery, which is
glued on.

Like cat no 12, this is a fine example of a work by
the Danzig amber and ivory worker, Christoph
Maucher.[1] In particular, the facial features of Minerva
recall those of Minerva, and even Mercury, in cat no
12. The dog is reminiscent of the dogs in the latter, and
in the Berlin *Judgement of Paris* attributed to Maucher.[2]
As in cat no 12, we see here the human interpretation of
a classical myth, each figure linked to the others by
gesture and touch. The carving is so close in style that it
is difficult to assign a different date, and it is likely that
the two pieces are roughly contemporary.

Rohde, p 43 and fig 145.
Verres, *Pantheon*, 1933, pp 244 ff.
W. Klein, 'Die Elfenbeinschnitzer-Familie Maucher
von Schwäb. Gmünd', *Gmünder Heimatblätter*, No 11,
6 Jahrgang, November 1933, pp 166–167.
Trusted, *Album*, III, fig 9.

Notes
1   See entry for cat no 12.
2   Berlin, Staatliche Preussischer Kulturbesitz,
    Skulpturensammlung, inv 859.

17   *Casket*

Probably Danzig: late 17th–early 18th century.
Amber; silver hinges.
Height 7.2 cm Length 13.1 cm Depth 10.5 cm
C.44-1923. Department of Sculpture.
Given by Mrs A.W. Hearn in 1923.

This casket is now in good condition, having been restored in 1983/4, with clear casting embedding resin used to strengthen cracked and broken amber panels.

This small casket rests on four square amber feet. Apart from its silver hinges, it is constructed entirely from clear amber panels glued and dowelled together. Six hexagonal amber panels, on the lid, four sides and base are incised with designs of harbour scenes with figures, and classical architectural features; each is flanked with panels incised with acanthus leaf designs. The thin style of the latter is typical of engravings of decorative ornament of the later 17th century.[1] The scenes are probably also derived from engraved sources, such as the work of J.W. Baur (c. 1600–1642) after Melchior Kuszel, or possibly Matthäus Merian the Elder (1593–1650) of Basle.[2] These were popular sources, and parallels can be seen in ambers from Danzig of the late 17th and early 18th century.[3] The toothed form of the hinges is characteristic of metalwork of the late 17th century,[4] while the form of the box was common in the late 17th and early 18th century, when small amber objects were made without a wooden support, recalling the early amber goblets and tankards of the late 16th and early 17th century, which were similarly mounted in precious metals.[5] Its fragility, and the impracticality of using it for storing objects which could damage it, imply that it was a late example of an object for a *Kunstkammer* collection.

Notes

1   Numerous ornamental engravings of engraved acanthus leaf of this type were published in the late 17th century, which were much used by German goldsmiths, eg Steinle, Indau, Unselt, Heckenauer etc. See E.A. Seemann, *Katalog der Ornamentstich-Sammlung des Kunstgewerk-Museums*, Leipzig, 1894, pp 20–22.

2   Cf one of the set of scenes engraved by Baur after Kuszel, *Allerhand Wunder-Würdige Meer und See Porten* (Victoria and Albert Museum, inv E.7621-1905) and a harbour scene engraved in 1622 by Merian. See L.H. Wüthrich, *Das Druckgraphische Werk von Matthaus Merian d.Ae.*, I, Basle, 1966, plate 331. I am very grateful to Anna Somers Cocks for her suggestions on sources.

3   Cf the landscape scene on cat no 18, and the Chinoiserie style of landscape seen on panels on the socle of cat no 12, and panels (now lost) from the amber throne from Danzig of 1677. See Baer, pp 122–124. An early 18th-century amber cabinet in the Herzog Anton Ulrich Museum Brunswick also has incised Chinoiserie landscape scenes. See Rohde, fig 293 (scenes not visible in this illustration).

4   Cf the hinges of some English and European boxes in the Victoria and Albert Museum, inv nos 408-1887, M700-1926 and M270-1960. I am grateful to Anthony North for his advice about hinges.

5   Cf a box in the Grünes Gewölbe, Dresden, inv III 88 ii, in G. Weinholz, *Gefässe und Geräte aus Bernstein*, Dresden, nd (after 1971) thought to be from Danzig and to date from the first half of the 18th century. See also Rohde, fig 303, for a similar example in Schwerin, and A. de Foelkersam, 'L'Ambre jaune et son application aux arts' (in Russian), *Staruie Ghodui*, XI, 1912, fig opposite plate 12, for a parallel example in the Hermitage, Leningrad, (also illustrated in Pelka, figs 62–64).

18 Cabinet

North East Germany, perhaps Danzig. Early 18th century.
Amber on wooden core. Metal lock and hinges. Later plastic and plaster
restorations.
Label underneath handwritten in ink reads: 'XVI (*sic*) cent. Amber Cabinet.
From Hon. W. F. B. Massey-Mainwaring sale, June 6th, 1898. 17th century
Italian work.' This was the sale held by Robinson and Fisher in London,
6–10 June 1898.
Height 42 cm Length 38 cm Depth 19 cm
C.706-1909. Department of Sculpture.
Given by Colonel F. Fearon Tipping in 1909.

Some amber panels are suffering from surface crazing and are cracked. Pieces of amber
are missing from the back and the protruding edges of the base and top, in some places
clumsily repaired with plastic. A clear amber panel is missing from the front of one of the
drawers. One of the six feet is missing. The original amber bouquets of flowers on the
top have been replaced by plaster coated with shellac. The wings of the figure of Cupid
on top are chipped.

This substantial two-doored cabinet is constructed from a mosaic of clear and cloudy panels applied to a wooden core (this is the incrustation technique; see also cat nos 9, 10 and 16), and is supported by five (once six) flattened feet, each one made of small panels of amber stuck on to a wooden core, with ivory strips applied to conceal the joins. The cabinet is surmounted by a figure group in clear amber of Venus and Cupid, and on the top of the sloping socle on which they are placed are four clear amber balls, one at each corner; five clear amber urns topped by bouquets of flowers (probably originally carved of white amber, now plaster replacements) are set below. On the front and sides of the cabinet some of the clear amber panels are carved on the underside with scenes and accompanying inscriptions, and laid over metal foil. They are difficult to decipher as the amber is suffering from surface crazing, and the foil deteriorated, but with a good light the subject matter can usually be made out. On the top left of the left-hand door (as we look at the cabinet) is a picture of a hand with a laurel wreath crowning a branch (?), with the motto, 'STET QUOCUNQUE LOCO' (it shall stand in every place). Below is a scene of arrows flying at a target by a tree, with the motto 'NULLI PENELLI BILE TELO' (no arrows (fired) in anger (land) on the target). On the top right of the same door a hand holds a laurel wreath and a sword lies nearby; in the distance is a ship; the motto is 'RECTO TENENTIS MERCES' (the reward of keeping to the right). Under this scene is one of a stork with a sword in the grasp of its foot, with the motto 'DUM VIGILAT SERVIT' (while he is awake, he serves).[1] On the right-hand door, at the top left, are three hands appearing from clouds holding swords, with the motto 'NON CEDAM MALIS' (I will not succumb to evil). Below this is a scene of a sarcophagus in a grove, the motto reading 'HUIUSQUE' (this man as well). On the top right of the same door a hand holding smoking torches (?) appears out of a cloud, while another pours water on an anchor; the motto reads, 'ABSQUE SPE VITA LABAT' (without hope life wavers); here Hope is symbolised by the anchor. Beneath this is a picture of a pyramid, with the inscription 'ALIO POENA AUT REMUNERATIO' (from another, punishment or reward). On the left side of the cabinet (as we look at it), are six further scenes with mottoes and, as on the front of the doors, they surround a larger oval central landscape without a motto. On the top left corner is a dove carrying an olive branch in its beak, flying over the Ark, and the motto 'MECUM PACEM FERO' (I bring peace with me); below this is a tree, with the motto 'FRUCTUS ET FLORES GRATI SUNT' (fruit and flowers are free), and at bottom is a scene of a ship at sea with a rising sun (with a smiling face!), and birds flying, and the motto 'INFAUSTUS ORIENS FUCAT' ('Red sky in the morning, shepherds' warning'). At the top right corner a hand holds a pair of scales, with the inscription 'PRINCIPIS ARCEI' (of a prince's citadel):[2] under this is a picture of a sunflower and the inscription 'SOLUS COR MEUM EST . . . ET (?) APERIT' (the sun is my heart . . . (illegible) and (?) it opens); at the bottom is a house among trees and a comet or shooting star, but no visible motto. On the other side of the cabinet are six more scenes again surrounding a larger oval central landscape without a motto. At the top left a helmet and two hands holding a laurel wreath and a sword appear out of a cloud, with the inscription 'IN MANU BELLI FINIS' (in the hand of war is the end); under this is a flower with the motto 'TEMPORE' (in time), and at the bottom swims a whale before a distant city, with the motto 'FELIX PARTUS' (happy bringing forth; presumably this is a reference to Jonah). On the other side, at the top corner is an island with ships and the motto 'ALOQUE DEFENDOQUE' (I both feed and defend); beneath, is a plant and the motto 'NEMO NE . . . IMPUNE' (nobody . . . (illegible) me with impunity), and at the bottom a fox is drinking at a pool with the motto 'MALO QUALE' (evil of whatever kind). These nineteen mottoes and illustrative scenes are in the tradition of 16th- and 17th-century emblem books, which were illustrated with engravings accompanied by aphoristic sayings exhorting the reader to live a good and virtuous life.[3] Some of the mottoes here have over-tones of folklore (such as the fox as a symbol of evil, and the observation that a red sky heralds bad weather), some of the Bible (such as the references to Jonah and Noah's Ark), and some seem to be relatively worldly advice, such as the allusions to war ('in the hand of war is the end', and 'no arrows fired in anger land on the target'). The combination of these relatively serious moral messages with the Goddess and God of Love portrayed on the top of the cabinet is not uncommon; a casket at Kassel has a relief of Venus but is surmounted with a *memento mori* of a putto with an hourglass,[4] and an apparently secular large standing cabinet in Dresden houses a crucifix.[5]

The doors open on to two sets of five drawers, with an archway between, and under this is a pentagonal shaped space, floored with lozenge-shaped panels of clear and cloudy amber, imitating a tiled floor, and walled with mirrors, with three cloudy amber half-pillars at the corners. This space might have housed a religious or secular statuette, such as the crucifix at Dresden (see above), or a small amber pot of carved flowers, as in another cabinet at Dresden.[6]

The drawers might have held writing implements, jewellery, or gaming pieces (a cabinet now in Berlin contains gamesboards in its base).[7] The inner faces of the doors are decorated with more clear amber panels carved on the underside, here placed over wood, not metal foil, giving a dark background to the lighter areas where the amber is carved; they depict fountains, trees and landscapes; one is an emblematic image of a bird on a pole (?) with arrows being fired at it. On the front of the drawers and pilasters on either side of the arch are amber panels over foil painted with floral forms. The classical profile bust of a helmeted soldier in a laurel wreath and flanked by two scrolls to form a pediment (all carved in white amber), above the arch accords with the classical Corinthian pilasters on either side.

The form of the cabinet recalls other early 18th-century cabinets probably from North East Germany, such as the two in Dresden,[8] one in Brunswick,[9] one at Gripsholm, Sweden,[10] one formerly in Berlin[11] and one recently acquired in Berlin;[12] the latter (perhaps made in Danzig) being particularly close to ours. An amber cabinet in Malbork is also broadly similar in form. This belonged to the Polish King, Stanislaw August, and its inscriptions (referring to historical events) indicate that it dates from after 1771; it too was probably made in Danzig.[13] The inventory of ambers formerly in the ducal collection at Brunswick records a cabinet very like this one. The description runs: '(A rectangular box made of different kinds of amber) decorated with carved views, on the lid a small group, Venus and Cupid, and on the four corners there are small flower pots, all of amber'.[14] It is tempting to identify this with C.706-1909, but there can be no certainty, as many such cabinets were being produced at this time.[15]

The inscription on paper glued to the underside of the cabinet, probably written in the late 19th century, is obviously mistaken in ascribing the work to 17th-century Italy.

Rohde, p 67, fig 300.

Notes

1 Compare Jacobus a Bruck Angermundt, *Emblemata Politica*, Cologne, 1618. Emblem XXXVII: An eagle grasps a sword in its foot, with the motto, 'DUM PROSIM' (while I may be of use).

2 I am very grateful to James Yorke for his advice on the translation of this phrase.

3 See M. Praz, *Studies in Seventeenth-Century Imagery*, Rome, 1964. See also A. Henkel and A. Schöne (Eds), *Emblemata Handbuch zur Zinnbildkunst des XVI und XVII Jahrhunderts*, Stuttgart, 1967. An amber cabinet in the Malbork Castle Museum, dating from the late 18th century has mottoes (unillustrated) in German inscribed on it. The cabinet is illustrated in Grabowska, although the mottoes remain unpublished. Cf also the mottoes on the gamesboard, cat no 3 and the cutlery handles cat no 25. Latin inscriptions are to be seen on a gamesboard in Kassel (see Pelka, pp 78–80), and on one in Uppsala, (see J. Svennung, 'De latinska inskrifterna på K. Vetenskapssocietetens bärnstensbrädspel', *Kungl. Vetenskapssocietetens Arsbok*, Uppsala, 1960, pp 123–138). The Uppsala ones are identified as stemming from Erasmus, *Catonis precepta*, *Mimi Publiani*, and *Septem Sapientum illustres sententie*.

4 Herzog, *Kostbarkeiten aus dem Landesmuseum, Kassel*, Stuttgart, 1979. Entry on *Bernsteinkästchen*.

5 Dresden, Grünes Gewölbe. Rohde, fig 297. It was given by Frederick William I to Augustus the Strong of Saxony in 1728, and is known to be from Königsberg, c. 1725, see Rohde, pp 66–67. See also Pelka, p 72; the cabinet housed various small courtly objects, such as small boxes, a gamesboard and gamespieces.

6 Dresden, Grünes Gewölbe. Rohde, fig 299. Thought to be from North East Germany, first half 18th century.

7 Kunstgewerbemuseum, Berlin, inv no 1980/60 (see note 12).

8 See notes 5 and 6.

9 Brunswick, Herzog Anton Ulrich-Museum. Pelka, fig 43. Rohde, figs 292–294. Reineking von Bock, figs 178, 179. Thought to be from Königsberg, c. 1720–1730.

10 From the Swedish Royal Collection, inv. HGK 410. This is signed "C.G. 1712".

11 Pelka, fig 42. Rohde, fig 291. Thought to be from Königsberg, early 18th century.

12 Berlin, Kunstgewerbemuseum, inv no 1980/60. Reineking von Bock, fig 201. Thought to be from Danzig, early 18th century. (Sold at Christie's, London, December 1978. Previously Ole Olsens Collection.)

13 See note 3. One of the inscriptions reads 'Der überfall des Königs Stantislaus Augustus zu Warschau Nov. 3 1771'. I am grateful to Elzbieta Mierzwinska for transcribing the inscriptions.

14 Entry no 127 from the unpublished inventory of 1770, housed at Brunswick, Herzog Anton Ulrich-Museum. I am grateful to Eliette Hulbert for her translation of the inventory.

15 For example, a cabinet sold at Sotheby's, London, 2 December 1969, lot no 82, corresponds closely to the description.

19   *Rosary*

South Germany, perhaps Gmünd. Probably 17th century.
Amber beads and silver filigree with two gilt medallions.
Length approx. 50 cm Diameter of each bead approx. 1.2 cm
155-1872. Department of Metalwork.
Bought from A. Pickert, Nuremberg, in 1872.

Some beads have been chipped or marked, and some have darkened in colour, but the
rosary is otherwise in good condition.

The rosary consists of seventy faceted clear amber beads: six decades of Ave-beads, six Pater-beads mounted in silver filigree, and a pendant of three Ave-beads with a Pater-bead mounted in silver filigree from which depends a cross and the large gilt medallion of the Virgin and Child. A gaud in the form of a smaller gilt medallion of the Virgin and Child is suspended in the middle of one of the decades.

Amber was used to make rosaries throughout medieval times and later:[1] they were listed in the royal collection of the Duke of Burgundy,[2] and some survive in collections today, such as that at Cluny,[3] the Museo degli Argenti in Florence,[4] and the Tradescant Collection housed in the Ashmolean Museum, Oxford.[5] The very nature of amber rosary beads, with their minimal carving, or turning, means that it is difficult to date them and locate their place of origin. Here, the six-decade form (which means that strictly speaking this is a 'chaplet' or 'corona', rather than a true rosary),[6] is particularly associated with Southern Germany,[7] and the silver filigree mounts accord with this. Goldsmiths of Gmünd in Swabia imported amber from at least the 16th to the 18th century in order to make rosaries with silver or silver-gilt filigree spheres.[8] It is the figurative style of the Virgin on the medallions which suggests that the rosary is 17th century. The Virgin is said to represent the Altötting Madonna;[9] this is a richly decorated 15th-century enamelled figure-group of the Virgin and Child with saints and donors, the Goldenes Rössel. Many votive rosaries have been offered at this shrine.[10] There does not seem however to be any direct connection between the Altötting Madonna and the one represented on these medallions,

but the association may have been suggested by the original recommendation to the Museum to purchase the rosary.[11]

S. Bury, *Jewellery Gallery Summary Catalogue*, London, 1982, p 242 no 7.

Notes

1 See Pelka, p 38 and Rohde, p 13. See also E. Wilkins, *The Rose-Garden Game, The Symbolic Background to the European prayer-beads*, London, 1969, p 46. I am grateful to Anna Somers Cocks for suggesting this book to me. A German rosary of the late 15th century of wooden beads with one amber gaud is in the Victoria and Albert Museum, inv 517-1903.

2 A. MacGregor (Ed), *Tradescant's Rarities*, Oxford, 1983, p 234, where the 1467 inventory of the Duke of Burgundy is said to list fifteen rosaries (Delabord, II, 1853).

3 Cluny Museum; inv no Cl 9528.

4 C. Piacenti Aschengreen (Ed), *Il Museo degli Argenti a Firenze*, Milan, 1968, cat no 523.

5 A. MacGregor (Ed), *op cit*, pp 234–235.

6 Wilkins, *op cit*, p 54.

7 *Ibid*, p 55.

8 *Ibid*, p 46 and plate 8 (3).

9 Bury, *op cit*, p 242.

10 See T. Müller and E. Steingräber, 'Die Französiche Goldemailplastik um 1400' in *Münchner Jahrbuch der Bildenden Kunst*, V (Dritte Folge) 1954, pp 39 ff. I am grateful to Marian Campbell for pointing out this article to me. See also Wilkins, *op cit*, p 30.

11 The Art Referee's Report by M.D. Wyatt of November 1871 (43253) is the only document known to refer to the rosary; this no longer survives.

20   *Head of a Woman mounted as a brooch*

Amber: German, perhaps Kassel; mid 18th century.
Brooch-pin and chain and separate pin: English; 1910–1922.
Amber, with gold brooch-pin, gilt brass safety chain and separate gilt brass pin
(in leather box).
Height 3.8 cm (approx) Depth 2.6 cm (approx)
A.46-1940. Department of Sculpture.
Bequeathed by Capt Walter Dasent in 1940.

The nose and part of the hair on the left shoulder are damaged, and there are surface
scratches and slight chips.

This carved amber head of a woman was probably originally a piece of small-scale sculpture, transformed into a brooch by the attachment of the gold brooch pin at the back in recent times (see below). The long hair is parted in the centre, and a plait (or perhaps wreath) is coiled around the head; this classical hairstyle, the oval face and Roman nose led to the work being considered antique (mounted in modern times as a brooch) when it was first acquired by the Museum. However, it is not Roman,[1] but more probably an imitation of the antique made in the 18th century. Two pairs of small holes drilled into the back of the neck, and a crescent-shaped roughened area of the amber here suggest that the head was originally supported on a socle. Although there are no exact stylistic parallels, small amber busts set on socles occasionally featured in *Kunstkammer* collections, and one, formerly in the collection of King Stanislaw August of Poland, was sold on the London art market in 1973.[2] The style of the carving of ours does not readily suggest any one artist, although it appears to be similar to works such as those produced in the circle of the amber and ivory sculptor Jacob Dobbermann (b. 1682; d. 1745), at the Court of Kassel in the early 18th century,[3] and considerably pre-dates the true neoclassical style.

The brooch is in a leather-box specially made for it, whose silk lining is printed with the names of the jewellers Longman & Strong i' th' arm, and Widdowson & Veale, of 1 Waterloo Place, Pall Mall SW1. Longman & Strong i' th' arm were first recorded as jewellers in 1875; they merged with Widdowson & Veale (an older firm) in 1910, but had moved from Waterloo Place to Albermarle Street by 1923. The gold brooch-pin was almost certainly made by them, and dates therefore from between 1910 and 1922.

Notes

1 Cf D.E. Strong, *Catalogue of Carved Amber in the Department of Greek and Roman Antiquities*, London, 1966, for a well-illustrated and extensive survey of carved ancient Roman ambers.

2 Christie's, London, 4 December 1973, lot 95, when it was bought by Edric Van Vredenburgh. It is now on the Paris art market.

3 Cf Reineking von Bock, figs 194, 195 (fig 196 erroneously illustrates two busts of agate, not amber), and the large bust of a woman in the Landesmuseum, Kassel (Rohde, fig 248 and Reineking von Bock, fig 189). Cf also the mid-18th century busts on the large amber chandelier at Rosenborg, in Copenhagen, designed by Marcus Tuscher, and made by Lorenz Spengler, c. 1753. See F.R. Friis, 'Den Store Rav Lysekrone paa Rosenborg', *Kulturhist Studier*, 1904/09, pp 58–61.

21   Casket

European, perhaps Venetian; 19th century (?), incorporating amber pieces
probably North East German; early 18th century.
Amber and ivory on wooden frame, metal hinges. Damask lining.
Height 33 cm Width (with feet) 63.5 cm Width (without feet) 51.5 cm
Depth (with feet) 46 cm Depth (without feet) 38 cm
Given by Mrs Ellen Hearn in 1923
C.19-1923. Department of Sculpture.

Some restoration has been carried out on the reliefs on the lid. There are signs that some
of the amber panels were originally from a different piece (see below).

This large casket with a hinged lid consists of amber and ivory panels and strips glued on to a wooden frame. The amber panels on the main body of the casket are carved in intaglio with landscape and figure-scenes. Other rhomboid-shaped amber panels with holes gouged out underneath are placed over gold coloured metal foil. The ivory panels are crudely incised with landscape vignettes, stylised acanthus leaves and other decorative motives, and the incisions darkened with ink. Amber swags of fruit hang from the base of the casket, while on the top are three amber oval scenes of nymphs and satyrs in deep relief. The casket rests on four elaborate feet; it is lined on the inside with red damask.

The uniqueness of the piece makes it difficult to classify easily. The most obvious anomaly is the placing of the intaglio amber panels. These are displayed with the carved side upwards, which means that the scene is less visible and less beautiful; it is highly unlikely that this would have been the original position. Indeed it is probable that the intaglio panels came from a different piece, and were united at a later date with the large wooden carcase, the clumsily incised ivory, and those amber panels which are plain and uncarved.[1] Two of the amber pieces have certainly been incorporated into a context for which they were not originally intended. These are the medallions, one on each side of the casket. The one on the right (as we look at the casket) shows Cupid carrying a lantern, with the motto (laterally inverted, as the intaglio, like all the others, is the wrong side up), 'Je re . . . plus celer' (I . . . (illegible) again more

quickly?). The other, on the left side, shows Cupid holding a ring. These medallions were once game-pieces; their style is typical of the early 18th century.[2]

The intaglio landscape scenes with their reminiscences of Chinoiserie probably derive from engravings, and recall the intaglio work on the early 18th century cabinet from Danzig, cat no 18.

The three oval carved medallions on the lid are weakly carved, but are also stylistically most likely to date from the early 18th century. They are bordered with faceted cubes of clear amber; this heavy feature, like the form of the extraordinary large feet, is an unattractive use of the material, and probably dates from the time the casket was constructed.

The main decorative elements are the intaglio carvings on the amber and ivory, the swags of fruit, and the figure carvings on the lid. Very little attempt has been made to show off the colour contrasts of amber; the overall form of the casket is unique in amber works, although it directly imitates wooden caskets. Early museum records show that when first acquired, the casket was thought to be Italian, and parallels can be cited with earlier wooden caskets from the Veneto.[3] The casket was probably made up from wood, ivory and pieces of amber in the 19th century, incorporating earlier pieces (the amber swags of fruit, intaglio, and low relief panels), which perhaps came from an 18th-century German casket broken beyond repair, along with two games-pieces. In its present form, it is a hefty monument to the decline of the art of amber working.

Notes

1   I am grateful to Waldemar Goralski for pointing out the comparative minimal amount of oxidisation on the surface of the plain amber panels (as opposed to the low relief and intaglio ones) which implies the former are of a much later date.

2   Amber games-pieces with inscriptions in French, the language of many European courts, dating from the early 18th century, survive at Cracow in the Czartoryski Collection, in the Grünes Gewölbe Collection at Dresden, the Nationalmuseum, Stockholm, and in the Landesmuseum Kassel.

3   See C. Alberici, *Il Mobile Veneto*, Milan, 1980, p 61, fig 76. This casket is said to be Venetian, second half of the 16th century.

22   *The Rest on the Flight into Egypt with the Miracle of the Palm*

Italian; late 17th century
Amber on lapis lazuli. Slate backing and gilt surround.
Height 27.5 cm   Width 21.2 cm
A.12-1950
Given by Dr W.L. Hildburgh FSA in 1950. Department of Sculpture.

The fingers of some angels and St. Joseph are missing, and the top of St. John the Baptist's cross is broken off. There have been minor breaks and restorations to the palm tree and amber background of the principal figure group; otherwise the piece is in good condition.

The subject of the scene is the miracle of the palm, which bent down to give its fruit to the Holy Family on their rest on the flight into Egypt. Angels then carried a branch up to Heaven to be planted there.[1] The Virgin and Child sit apparently in the lap of St. Joseph recalling the usual grouping of the Virgin and Child with St. Anne;[2] the infant St. John kneels at their feet. Three angels kneel adoring, one offering fruit which has been picked from the palm on the left, while four others hover overhead. The central group of the Virgin, St. Joseph, the Christ Child and St. John the Baptist is carved from one lump of amber; the two angels on their left are also carved from one piece; possibly originally all six figures were made from the same piece which was later broken and restored; the other figures and landscape features are separate. The amber has been fixed on to the lapis lazuli backing; a composition in deep relief fixed in such a way seems to have been a rare and skilful use of the material. An example probably roughly contemporary with this one, depicting the Christ Child and the Infant Baptist with a lamb, probably carved in Kassel, is in the Kassel Landesmuseum.[3] Unlike ours, it is backed on to wood, with amber of contrasting colours forming a cloud surround and angels' heads. The opulent use of lapis lazuli for the backing shows how valuable a commodity amber was considered to be. An even closer parallel with the present relief, now lacking its original backing, is in the Royal Scottish Museum, Edinburgh.[4] It depicts the Baptism, in a style strikingly close to A.12-1950, and must be from the same workshop. It too has a rounded top, and although

The Baptism, reproduced by kind permission of
the Royal Scottish Museum, Edinburgh.

now has a later elaborate carved wooden frame, may once have been mounted in gilt metal like ours. It is however considerably smaller in scale (Ht 22 cm, W 15.5 cm), and is therefore unlikely to have been a pendant to this one. A clue is provided to the origin of both by the inscription on the Edinburgh relief; written in ink on the base of the frame is the word 'Batista', indicating that at least at one time it was in Italy. The style of the two pieces, probably derived from prints of Italian High Renaissance or Mannerist paintings would certainly support this.[5] The shape and scale of both works are similar to the relief of the *Adoration of the Shepherds* which forms the focal point of the Museum's monumental amber altar (cat no 9). Although the *Rest on the Flight into Egypt* and the *Baptism* would have been inappropriate subjects for the central reliefs on an altar, they could possibly have been used as subsidiary reliefs on a large one. Amber was traded with Italy, and these two reliefs may have been the products of a workshop, not necessarily exclusively an amber workshop, in Rome,[6] or conceivably Sicily, where the tradition of coral carving provided a parallel art–form.[7]

Notes

1 See G. Richter, *Ikonographie der Christlichen Kunst*, I, Gütersloh, 1966, pp 128–129.

2 I am grateful to Dr Jennifer Montagu for suggesting this.

3 Kassel, Staatliche Kunstsammlungen, inv no B.VI.15 Rohde, fig 249.

4 Royal Scottish Museum, Edinburgh, inv no 1869.2b.8.

5 No precise source has been identified, but they are close in style to the engraved work of Domenico Pellegrino, called Tibaldi (1527–1596); see *Bartsch*, XXXIX, p. 27.

6 Cf the work by Francois Duquesnoy, probably carved while he was in Rome. See A. Kosegarten, 'Eine Kleinplastik aus Bernstein von Francois du Quesnoy', *Pantheon*, XXI, 1963, pp 101–108. An unpublished relief of the *Adoration of the Shepherds*, probably Italian, 17th century, is housed in the Capodimonte Gallery, Naples.

7 A close parallel is a coral carving of the Immaculata, sold at Sotheby's, London, 13 November 1975, lot 177. A tradition of folk-carving of Sicilian amber existed in Sicily, probably dating back to the 17th century, but the works have subsequently been dispersed. I am grateful to Helen Fraquet for showing me Rosalind Denny's unpublished notes and photographs of amber groups formerly in Palazzo Atenasio, Taormina. Cf also the sale of the Kitson Collection at Sotheby's, London, 23 January 1961, lots 99, 145, 153.

23 Head of a Saint

Italian (?); late 17th–early 18th century.
Amber
Height 7 cm   Width 5.5 cm
A.13-1950. Department of Sculpture.
Given by Dr W.L. Hildburgh FSA in 1950.

Scratch marks appear around the edge of this medallion, and there is a little surface crazing, but otherwise the piece is in good condition.

The scratch marks around the edges and a small hole drilled at the top of this clear red amber medallion suggest that it was once set in metal mounts and worn as a devotional pendant, or was mounted as a standing ornament, or on the lid of a box.[1] The lack of severe oxidisation of the surface means that the luminous qualities of the amber and resemblance to semi-precious stones are still evident. The green fluorescence visible at the back might suggest that the amber is Sicilian, rather than the more usual Baltic amber; however, such fluorescence is not conclusive, and as the trading of Sicilian amber seems to have been minimal at this time, it is more likely that this is of Baltic amber.[2] The subject, the head of an unidentified saint, with long flowing hair and beard, his head turned to the right, and his mouth slightly open, is finely carved in relief, but cannot be readily assigned to a particular district, or even country. Its dating is also problematic: it probably derives from a painting, or a print of a painting of the Mannerist or Baroque era, and is closer in style to the Italian than Northern Schools. Along with the relief of the Holy Family (cat no 22), this is therefore probably one of the rare Italian ambers in the collection, and dates from the late 17th to the early 18th century.

Notes

1 Cf the box, standing pyx, and pendant (all now destroyed or lost) illustrated in Rohde, figs 60–62.

2 The fluorescence is caused by impurities in the amber. See B. Kosmowska-Ceranowicz, 'La Struttura dell'Ambra. La Ricchezza delle Varieta' in *Ambra oro del Nord*, Venice, 1978 (Exhibition catalogue).

Cutlery and Flatware

Cutlery and flatware handles were probably manufactured in Königsberg, Stolp, Elbing and Danzig, as well as in London. Few exact dates or clues to places of origins survive, and the identifications given are partly based on the comparison of technical methods used in other ambers known to be from North East German cities. However, techniques could no doubt be learnt, and the Prussian and Pomeranian cities were close enough to each other for information to be easily transmitted, while Continental craftsmen such as Peter Spitser (see cat no 36) are known to have settled in London, and would have imported techniques. The marks on the blades, when they have been identified, are not always conclusive, as blades could be fitted to handles made elsewhere, and the hafter was anyway a different craftsman from the cutler.[1]

Notes

[1] See Blair, p 444, Beard, pp 96–97, Hayward, pp 164 and 167. See also introduction.

24 Three Knives and one Fork in a Tooled Leather Case

North East Germany, probably Königsberg. Late 16th century.
Bands of ivory, ebony and amber with brass washers and inset studs of coloured
pastes, imitating precious stones. Steel blades and ferrules. Stamped leather case.
Length of handles 6.9 cm, 7.1 cm, 7.8 cm, 6.8 cm
Length of knives 17.9 cm, 19.2 cm, 20.1 cm, 17.4 cm
Length of handle of fork 6.8 cm
Length of fork 18.4 cm,
Height of case 19.8 cm (without lid)
524 to E-1893. Department of Metalwork.
Bought from Mr F.E. Whelan; Bateman Sale (Part I) at Sotheby's London,
14 April 1893, Lot 43.

The handle of the fork has lost its original amber band; the ivory and amber of the other
handles are scratched. The paste studs probably replace original amber ones (cf cat no 25,
see below). The tooled leather case is damaged on the surface in several places, and its lid
is missing.

This fork and the three knives belong to a set of eleven, which were split up shortly after the Museum acquired them, when three knives were sent to the National Museum of Ireland in Dublin, three to the Royal Scottish Museum, Edinburgh, and one to the Sheffield City Museum and Art Gallery. As early as 1389, amber handles for cutlery were recorded (see Introduction), and this set therefore represents a tradition dating back to medieval times. According to the records of 1893, on the tooled leather case was inscribed repeatedly 'VERBUM DOMINI MANET IM ETO' (The word of God remains for eternity) but this was only on the lost lid, as no inscription is now visible on the case, although letters can be seen on the lid in a photograph of 1961. (See illustration). Although dated by Bailey, Venetian 17th century,[1] the design of the fork and knives is similar to cutlery made in North East Germany at the end of the 16th century,[2] if it is assumed that the present studs of coloured paste were originally amber. The marks on the blades have not been identified. ?

Bailey, fig 39.

Notes

1 Bailey, p viii. They are described as Venetian 16th century in the Sotheby's Sale Catalogue of 1893.

2 Cf the knives at the Luitpold Museum, Würzburg, dated 1588. Rohde, fig 41, Reineking von Bock, fig 80.

25 *Presentoir and Serving Fork*

Königsberg; c. 1600.
Amber and ivory. Steel tines and blade.
Length of handle of fork 13.2 cm Length of whole 43 cm
Length of handle of presentoir 13.5 cm Length of whole 53.3 cm
M.920 and A-1926. Department of Metalwork.
Bequeathed by Lt Col G.B. Croft Lyons FSA in 1926.

Some amber panels are missing, and elsewhere some are cracked or suffering from surface crazing.

The presentoir, a spatula-like knife, was introduced in the Middle Ages to serve meat,[1] and is here paired with a serving fork. The elaborate handles are ivory inlaid with amber, with a metal tang running up through the centre of each surmounted by a gilt ball finial. Clear amber discs are set into the cusped ivory forms at the heads of the handles, the amber being placed over gold-coloured foil to accentuate its reflective qualities (some of the amber discs and foil are missing on the handle of the presentoir).

The central section of the presentoir is decorated with, on one side, *St. George Killing the Dragon*, and on the other, *Mettius Curtius leaping into the Gulf*. These are both painted on foil which is viewed through clear amber. On the sides of this section are two panels of lettering (also painted on foil under clear amber): 'AMICICIA' (Amicitia; with friendship), and 'ET EXTERIVS' (on the outside too).[2] The corresponding images on the fork are difficult to read because of the severe surface crazing of the amber, but one is *David and Goliath*[3] and the other, the *Return of the Two Spies Joshua and Caleb with Grapes from the Promised Land.*[4] The two panels of writing are 'DISCE VIVERE' (learn to live), and 'DISCE MORI' (learn to die). Bands of ivory inset with clear amber discs over foil run round the handles; they are incised with dots and crescent lines darkened with ink. The inset discs and cusped handles are typical of those produced in Königsberg in the late 16th and early 17th century; parallel examples are to be found in Cracow, Stuttgart, Dresden and Munich.[5] These handles, like ours, are all likely to be Königsberg ones of the early 17th century. The mark on the blade of the presentoir has not been identified.

Trusted, *Album*, fig 1.

Notes
1 See *Masterpieces*, p xii.
2 I am grateful to James Yorke for advice on the translation of the Latin.
3 *I Samuel*, XVII.
4 *Numbers*, XIII, v 23. I am grateful to Jonathan Voak and Father Anthony Couchman for locating this reference.
5 The Czartoryski Collection, Cracow, the Grünes Gewölbe, Dresden, the Schlossmuseum, Stuttgart (Rohde, figs 37–40, Reineking von Bock, fig 76). Handles in the Bayerisches Nationalmuseum, Munich are also similar (inv R1295, R1296, unpublished). Cf also a carving knife in the Howard E. Smith collection of cutlery; see Hayward, fig 7.

26   Carving Knife

Königsberg, early 17th century.
Amber and ivory; brass washers; gilt steel ferrule and steel blade; gilt brass cap,
brass finial.
Length of handle 12 cm Length of whole 35.1 cm
M.95-1923. Department of Metalwork.
Given by Mrs A.W. Hearn in 1923.

There is some damage: the amber is in places chipped, and suffering from surface crazing.

The hexagonal handle of this knife is decorated in a
similar way to cat no 25. Here, at the top and bottom of
the shaft of the handle ivory panels, in the form of
darts, like the points on a backgammon board, are inset
with discs of amber over foil. These alternate with
corresponding inverted darts of clear amber panels
over arabesque designs painted on foil. Between these
two bands of triangular designs, in the middle section
of the handle, are six clear amber panels over foil
painted with full-length figures perhaps representing
different crafts. The handle is connected to the gilt
ferrule of the sharply tapering steel blade by a steel tang
running through the centre, with a brass ball finial. The
cap is of gilt brass with a clear amber hexagonal panel
over foil painted with arabesque designs, forming the
transition between the finial and the shaft. It is edged
by a brass washer. Two horizontal bands of clear
amber with brass washers separate the central
figurative section from the bands decorated with darts
of ivory and amber.

Parallels with similar handles suggests that this one
dates from the early 17th century and comes from
Königsberg.[1] The marks on the blade have not been
identified. ✦

Bailey, fig 46.

Notes
1   Cf the handles in the Czartoryski Collection, Cracow thought
to be from Königsberg, c. 1600 (Rohde, fig 40). Cf also a knife
and fork in Sheffield, City Museum and Art Gallery, inv
L1931-3 (unpublished).

27 Three Knives and one Fork

North East Germany, perhaps Stolp. Dated 1634.
Amber over painted metal foil; silver panels, gilt steel ferrules and steel blades
and tines. Gilt brass caps and finials.
Length of handles of knives 9.1 cm Length of whole 25 cm
Length of handle of fork 9.8 cm Length of whole 22.6 cm
1391, 1391B, 1391H, 1391K-1888. Department of Metalwork.

Part of a set of twelve knives and one fork bought in the sale of the collection of the Earl
of Londesborough at Christies, London, 10 July 1888, lot 665. Some of the amber is
chipped, but generally these objects are in good condition.

The handles of all four pieces are of the same form: a steel tang, with gilt brass cap, and ball finial runs down through the hollow centre of the amber to attach it to the gilt steel ferrule and steel blade or tines. A piece of clear amber carved in the shape of a segmented vase forms the top of the handle, while the main body of the handle consists of a clear amber panel over foil painted in red, black and gold with a full length figure of an Apostle or Christ, and inscriptions with extracts in Latin from the Gospel of St. John and the Apostles' Creed.[1] On the sides of each are set two silver panels inscribed 'Leonardus Marius' and '1634'. This probably refers to the owner rather than the maker.[2] The knives and form originally belonged to a set of twelve knives (the Apostles) and one fork (Christ); all were housed in a leather case. The other knives and the leather case have subsequently been lost or destroyed, and only these now remain. The fork handle has a painted figure of Christ and the inscriptions 'SALVATOR' (The Saviour), 'EGO SUM VIA VERITAS ET VITA' (I am the way the truth and the life).[3]

1391B-1888 has the figure of St. Matthew and the inscriptions 'ST MATTHAEUS' and 'INDE VENTURUS EST IUDICARE VIVOS ET MORTUOS' (thence He will come to judge the quick and the dead). 1391H-1888 has the figure of St. James the Less and the inscriptions 's. JACOBUS MINOR' and 'CREDO IN SPIRITUM SANCTUM SANCTAM ECCLESIAM CATHOLICAM' (I believe in the Holy Spirit, the Holy Catholic Church). 1391K-1888 has the figure of St. Peter, and the inscriptions 'ST. PETRUS' and 'CREDO IN DEUM PATREM OMNI POTENTUM' (I believe in God the Father Almighty). The marks on the blades, a cross and a crescent, are identical with those on a knife and a fork in the Malbork Castle Museum, those on cat no 28 and on a knife in the James A. de Rothschild Collection at Waddesdon Manor,[4] but they have not yet been identified. Handles of a similar, though not identical form, exist in other collections.[5] It is difficult to assign a specific place of origin, but they may come from Stolp, where this type of handle was made.

Bailey, p 13 and fig 45.

Notes

1   For the Apostles' Creed see *New Catholic Encyclopaedia*, IV, New York, 1967, p 436.

2   Cf cat no 35.

3   The Gospel according to St. John, XIV, vi.

4   Inv no M2M/b/967, illustrated in Grabowska. See Blair, pp 438–439. The knife-handle is attributed tentatively to Königsberg, although the author states that similar handles may have been made in Danzig and Elbing (Stolp is not mentioned).

5   See Rohde, figs 45, 46, 153, and Reineking von Bock, figs 103–107.

28   *Wedding Knife and Fork*

North East German, perhaps Stolp; mid 17th century.
Amber handles; steel blade and tines. Encrusted silver gilt ferrules. Gilt brass cap.
Length of handle of knife 7.1 cm Length of whole 22.8 cm
Length of handle of fork 6.7 cm Length of whole 20.3 cm
M.99 and A-1923. Department of Metalwork.
Given by Mrs A.W. Hearn in 1923.

There are some marks and chips; the knife handle is more badly damaged than the fork.

These clear amber handles are carved with the heads of a man (on the knife) and a woman. The lengths of the handles are carved in low relief with bands of cabling. Steel tangs covered in gold-coloured foil run through the centre of each, with a gilt brass cap finial. The gilt steel ferrules are encrusted in silver gilt with a pattern of stylised scallops. These are analogous to those of cat nos 35 and 36, but in this case they are German rather than English. The hairstyles and collars of the figures imply that they date from the mid 17th century; analogous examples can be seen in Cologne.[1] Although formerly thought to be from South Germany,[2] the handles were almost certainly made in a city with a tradition of amber carving in North East Germany; the similarity with handles known to be from Stolp in Pomerania means that this could be their place of origin.[3] The marks on the blade, a crescent above a cross on its side are similar to those on cat no 27, to a knife and fork in Malbork, and to a knife in the James A. de Rothschild Collection at Waddesdon Manor.[4]

Bailey, p 13 and fig 47.

Notes
1   See Reineking von Bock, fig 78. Cf also Rohde, figs 45, 46, 154. See also cat no 29.

2   Bailey, p 13.

3   See Rohde, pp 27–28.

4   Blair, pp 438–439. See note 4 to cat no 27.

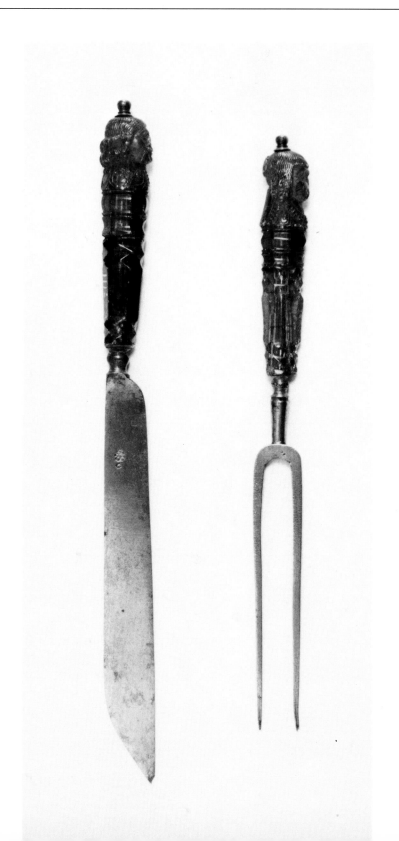

29   *Wedding Knife and Fork*

North East Germany, perhaps Stolp. Mid 17th century.
Amber; gilt steel tang and ferrule; brass finial.
Length of knife 21.7 cm (whole) 7.6 cm (handle)
Length of fork 18.6 cm (whole) 7.1 cm (handle)
111 and A-1872. Department of Metalwork.
Bought from S. Rosenau, Kissingen in 1872.

The faces of both busts have been badly chipped; otherwise they are in good condition. When purchased they were in a stamped and gilt leather case, subsequently lost.

This knife and fork, with the bust of a man carved in clear amber on the knife, and a woman on the fork could have been a wedding-gift.[1] A steel tang with a brass finial runs from the gilt steel ferrule of the steel blade and tines respectively through the centre of each of the handles. The handle of the fork appears to be one piece of amber, the knife two fitted together. Similar examples exist in the Victoria and Albert Museum and elsewhere.[2] The mark on the blade of the knife has not been identified. Such handles were made throughout the 17th century, and it is impossible to assign a specific date or location to them, although, like cat no 28, these may come from Stolp. The hairstyles and collars of both figures implies a date from the mid 17th century.[3]

Notes
1   Cf Bailey, p 4.
2   Cf, cat no 28, and Rohde, figs 45, 46, 154 and Reineking von Bock, fig 78.
3   Cf the representations in amber of the Great Elector and his consort, Rohde, fig 144.

30 Fork

North East Germany. 17th century (?) Head perhaps 19th century.
Amber. Silver ferrule and tines. Silver cap.
Length of handle 5.6 cm Length of whole 14.8 cm
1060-1902. Department of Metalwork.
Given by J.H. Fitzhenry, as part of the Fitzhenry Collection, in 1902.

The head may have been added on to the main body of the handle at a later time (see below). The amber is chipped, and the head is particularly badly damaged.

The amber handle consists of a female head broadly carved, placed on top of a block of amber incised with a tendril design. A silver tang runs through the middle of the handle to attach it to the ferrule. The head, topped by a silver cap, fits poorly on to the rest of the handle, and could well be a later addition replacing a broken top. The ferrule is marked with a dolphin, a Dutch duty-stamp in use between 1859 and 1893.[1] The tendril designs suggest a date during the late 17th century.[2] The simple carving means that it is not possible to give an exact location, although it is probably from North East Germany.[3] The head may have been a replacement added in the 19th century.

Notes

1 See E. Voet, *Nederlandse Goud- en Zilvermarken*, The Hague, 1975, p 47, fig 2.

2 Cf the designs on cat no 11.

3 Cf the handles in the Kunstgewerbemuseum, Cologne, inv K31, thought to be c. 1680–1690 and to come from Stolp. Reineking von Bock, fig 78.

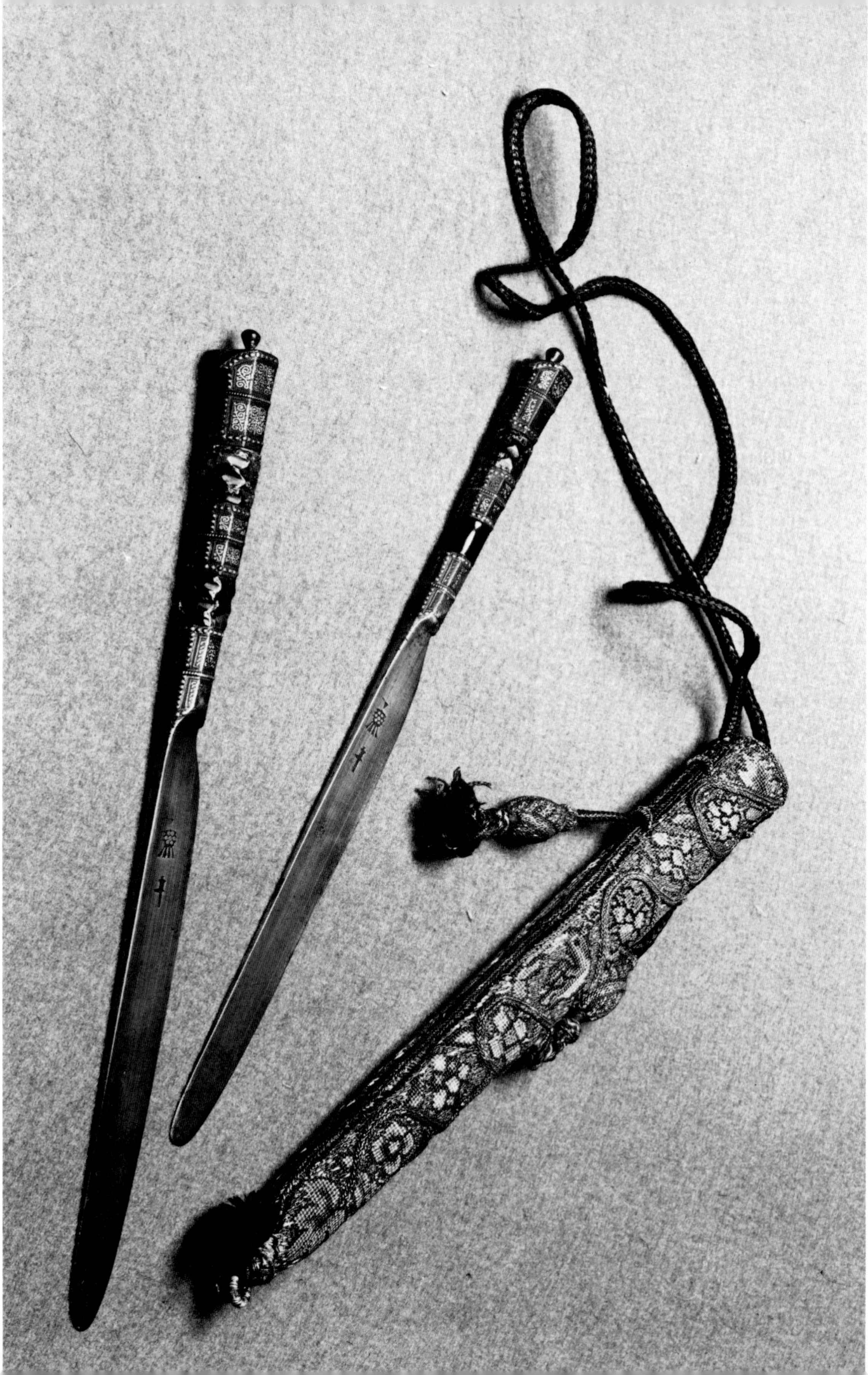

33 *Pair of Wedding Knives and Case*

English (London); about 1610–1620.
Mark of John Jenkes (active 1606/7; d. 1620) of London.
Amber, steel damascened in silver and gold.
Case: silver-gilt and silk threads worked on to canvas. Carrying-string of silk braid.
Length of handles 8.2 cm Length of whole 20.3 cm
T.55 to B-1954. Department of Metalwork.
Bequeathed by Sir Frederick Richmond, Bart in 1954.

The knives and case are in good condition.

The faceted clear amber panels on the knives form two bands on each, alternating with steel sections damascened in silver and gold with raised dots and arabesque designs. The style of the knives indicates that they are probably early 17th century, and the marks on the blades, a dagger and thistle, are those of John Jenkes (active 1606/7; d. 1620).[1] A similar pair of knives, dated 1610, also with John Jenkes' marks was formerly in the Doucean collection at Goodrich Court.[2] Like cat nos 34–36, these are examples of amber worked in England, although the simple faceting is less sophisticated than the more elaborate decoration seen in them. The embroidered case would have been worn by the owner at her waist.[3]

Notes

1 See Welch, II, p 41. See also Beard, pp 94–95.

2 See S.R. Meyrick, 'Catalogue of the Doucean Museum at Goodrich Court, Herefordshire', *Gentleman's Magazine*, VI, October, 1836, p 382, no 8 (Clive Wainwright kindly drew my attention to this catalogue) and Beard, p 93 and fig ii. Cf also knives with amber handles by John Jenkes' son, Joseph, dated 1633, in the Sheffield City Museum and Art Gallery, inv L.1932-5.

3 See *Masterpieces*, p 8. See also Beard, pp 91–97, and Hughes, pp 666–667.

34 *Knife*

English; London (?); early 17th century (?).
Handle: Amber, mother o' pearl, ebony, brass washers.
Blade: Steel with traces of gold inlay. Perhaps by John Arnold (active 1606–1607)
Length of whole 17.2 cm Length of handle 7.2 cm
539-1901. Department of Metalwork.
Bought from C.H. Shoppee Esq in 1901.

This handle is decorated with clear amber panels, chequerwork of mother o' pearl, bands of ebony and brass washers. The mark on the blade, a pair of tongs, accords with marks granted to several London cutlers from the early 17th century up to 1675.[1] There are no obvious parallels stylistically with German handles, but they resemble cat no 33 and another pair of knives with handles of a similar form, which are known to be English, dated 1623 and were exhibited in 1979.[2] The amber for this knife was probably imported as a raw material and worked by the hafter.[3]

Notes

1   The mark was granted to John Arnold on 15 January 1606–7, William Savage on 10 March 1636–7 and 25 March 1664, William Boswell on 7 August 1666 and Thomas Elliot on 9 February 1668–9, William Bosveile's son on 18 February, 1673–4, and Jarvis Boswell on 23 September 1675. See Welch, II, pp 41–42.

2   See *Masterpieces*, cat no 24A, illustration 2, and p 8. These have bands of amber alternating with sections of damascened steel.

3   Amber was imported to England by at least 1656. See introduction and note 4 to entry for cat no 35. See also Hayward, p 164, and Beard, p 96.

35   *Pair of Wedding Knives in Embroidered Case*

English; London (?); dated 1638.
Amber and ivory handles; silver filigree finials; silver encrusted ferrules; steel blades. Embroidered silk case.
Length of handles 8.8 cm Length of whole 29.8 cm
M.12 to B-1950. Department of Metalwork.
Bought at Christie's, London, 1 May 1950, lot 143.

Both handles are in good condition, although the amber panels are slightly chipped and are suffering from some surface crazing.

The finials of these octagonal handles are silver filigree, above convex drop-shaped clear amber panels over white amber profile busts of a man and a woman respectively, set against foil.[1] A steel tang covered in gilt foil can be seen through the clear amber panels around the handle connecting it to the silver encrusted ferrules. In the central section clear amber panels over painted foil alternate with ivory strips. The foil is decorated with designs of coursing dogs and other animals, the name 'ANNA MICKLETHWAIT', and the date 'ANNO 1638'. The name is probably that of the owner for whom the knives were made on the occasion of her wedding; she would have worn them in the embroidered case.[2] They are very close to a pair also dated 1638 (made for 'TABATHA GELDART') decorated in a similar way, now owned by the London Cutlers' Company.[3]

The usual assumption that works of amber come from the Baltic coast is here brought into question. Although it is possible that the handles were commissioned abroad and brought back to England, other evidence suggests that occasionally amber was imported and manufactured in London, although the techniques must have been learnt from Baltic craftsmen.[4]

Trusted, *Album*, figs 2 and 3.

Notes

1 Cf the medallion of King Stefan Batorys of Poland, illustrated in Grabowska. See note 1 to cat no 3.

2 See *Masterpieces*, p. 8. See also Blair, p 443.

3 *Masterpieces*, cat no 25, fig 5. The central sections are like those on a pair of wedding knives of c. 1610 in the Museum of London. See Hughes, p 666.

4 See Introduction, and Blair, p 444. See also Beard, pp 96–97. Apart from these two pairs of handles, evidently made for English customers, see cat no 33.

36 Pair of Wedding Knives and Case

Handles: London; dated 1639. Blades by Peter Spitser of London (active 1621 onwards).
Amber and ivory; steel blades. Gilt brass finials. Silver encrusted ferrules. Case: metal thread on a silk core couched onto a silk velvet ground, with leather lining.
Length of handle of M.444 6.5 cm Length of whole 20.4 cm
Length of handle of M.444A 6.2 cm Length of whole 19.9 cm
M.444 and A-1927. Department of Metalwork.
Bought from Miss E.M. Ashworth in 1927.

Some amber panels are missing. The original velvet case for the knives is much dilapidated, and the metal thread worn and tarnished.

The handles are of ivory overlaid with slender oblong clear amber panels, painted on the underside, over foil. The steel tangs are surmounted by a gilt brass ball finial, and the ferrules are encrusted in silver with a pattern of fruit-like forms. Designs of coursing dogs and hares and the date 1639 are painted on the underside of the amber panels of the two handles.

On M.444-1927 are the additional words 'ANNO' and the name (of the owner) 'AYLLS'. The other knife was probably similarly inscribed, but now only remnants of 'ANNO' remain, and what was the name panel is missing. The bare foil under the missing panels shows that here the lettering and animals are painted on the underside of the amber, not on the foil.[1] In general the type of decoration corresponds with work produced at Königsberg,[2] but like cat nos 33–35, these may have been made in London;[3] the marks on the blade, a unicorn's head and a dagger, are those of Peter Spitser, who was a German immigrant working in London; the mark was registered on 7 August 1621.[4]

Beard, fig IX, p 97.

Notes

1 Cf numerous other ambers where the foil is painted and placed under clear amber, eg cat nos 3, 25 and 27. A near-identical effect is achieved, but the means are different.

2 Cf ambers cited in note 1.

3 See note 4 to cat no 35.

4 Welch, II, p 42.

Appendix Chinese Ambers in the Far Eastern Department (Compiled by Craig Clunas)

Necklace
Amber, jadeite and rose quartz beads
Middle of the 19th century
Traditional unsubstantiated provenance to the
Summer Palace, Peking
1126-1874

Miniature Vase
Amber
19th century
Height: 13 cm
From the Museum of Practical Geology, Jermyn
Street
5544-1901

'Buddha's Hand'
An inedible citrus fruit used as a decoration
Amber or copal
18th–19th century
Length: 12.9 cm
Bequeathed by W.H. Cope Esq
725 & A-1903

Vase and Waterpots
In the form of a fruiting peach spray, used as a
writing accessory
Amber
18th century
Height: 15.5 cm
Bequeathed by W.H. Cope Esq
748 to C-1903

Snuff Bottle
Amber
19th century
Height: 5.1 cm
Salting bequest
C.1581-1910

Snuff Bottle
Carved in low relief with flowering shrubs
Amber
19th century
Height: 9 cm
Salting bequest
C.1857-1910

Saucer
Carved in the form of a lotus leaf
Amber
18th–19th century
Diameter: 9.3 cm
Salting bequest
C.1893-1910

Saucer or Waterpot
Carved in the form of a hibiscus flower with leaves
Amber
18th–19th century
Diameter: 9.5 cm
Salting bequest
C.1901-1910

Carving
One side showing a dragon among clouds, the other
with a carp among waves
Amber
19th century
Length: 7.6 cm
Given by Mrs M.E.A. Wallis in memory of Edwin
Harold Ellis
A.1-1964

Index